CANYONEERING

BEGINNING TO ADVANCED TECHNIQUES

CHRISTOPHER VAN TILBURG, M.D.

THE
MOUNTAINEERS

Published by
The Mountaineers
1001 SW Klickitat Way, Suite 201
Seattle, WA 98134

First edition, 2000

Published simultaneously in Great Britain by Cordee, 3a DeMontfort Street, Leicester, England, LE1 7HD

Manufactured in the United States of America

Managing Editor: Kathleen Cubley
Editor: Uma Kukathas
Cover design, book design, and figure illustrations: Ani Rucki

All photographs © 2000 by Christopher Van Tilburg unless otherwise indicated.

Cover photographs: *Dirty Devil Canyons, southern Utah* © James Kay. Insets, back cover: left, *Natural bridge in canyon country, Escalante, Utah,* right, *Trail's end, Escalante, Utah.*

Frontispiece: *Squeezing sideways through the slot, Robbers Roost, Utah*

Library of Congress Cataloging-in-Publication Data
Van Tilburg, Christopher.
    Canyoneering : beginning to advanced techniques / Christopher Van Tilburg.—1st ed.
        p.   cm.
    Includes bibliographical references and index.
    ISBN 0-89886-704-5 (pbk.)
    1. Canyoneering. I. Title.
GV200.35.V25 1999
796.51—dc21   99-050610
                                                                                    CIP

796.51
VAN
7/00

FOR SKYLAR

Contents

Acknowledgments

This project was not a solo task, and I am grateful for the time and contributions of many people. Jim Blanchard, Jon Ciambotti, Matt Dinsmore, Geoff Richardson, Matt Ryan, and Peter Van Tilburg provided input on the manuscript. Leki Poles, Okespor, Gregory Mountain Products, Patagonia, and Tubbs Snowshoes provided equipment and clothing. Jim and Susie Kay provided a cover shot and an abundance of canyon adventure. Thanks to Erik Benner and Eric Idhe for contributing photographs. Rich Carlson allowed use of the American Canyoneering Association rating system. Fellow Wilderness Medical Society members Steve Bezruchka, Karl Naumann, and Bill Robinson gave me opportunity to promote canyoneering safety at the society's conferences and in its publications. My parents, Wayne and Eleanor Van Tilburg, support me continuously and indefinitely.

The crew at The Mountaineers Books have been wonderful. I especially thank Helen Cherullo, Kathleen Cubley, Margaret Foster, Art Freeman, Emily Kerr, Janet Kimball, Alison Koop, Uma Kukathas, Ani Rucki, Alan Stark, and Margaret Sullivan.

Most of all I thank my partner, Jennifer Wilson, who helped save some of the canyon country for all of us.

PLANNING AND PREPARATION

1

INTRODUCTION

Canyoneering, or canyoning as it is sometimes called outside North America, is, by its simplest definition, traveling through canyons. This relatively new sport borrows skills and techniques from hiking, backpacking, mountaineering, caving, and rock climbing. Canyoneering usually refers to hiking on foot, but some canyoneers use mountain bikes and others use watercraft such as rafts, canoes, or kayaks.

Basic canyoneering involves hiking or trekking through canyons, either upstream or downstream. The skills and equipment required are similar to those for backpacking, with some important modifications. Advanced trips usually entail wading across streams and scrambling over boulders or downed logs. Often this involves walking through water, swimming through pools, fording streams, negotiating waterfalls, squeezing through narrow passages, and climbing over log jams, rocks, or large boulders.

Technical canyoneering can entail rappelling into dank caverns, swimming through deep pools, rock climbing up or down cliffs or dryfalls, descending or ascending waterfalls, avoiding flash floods and high water, and traveling long stretches with little or no water. Like caving and mountaineering, technical canyoneering requires proficiency in a wide range of wilderness techniques, including rappelling, rock climbing, swimming, minimal camping, advanced weather reading, and difficult routefinding.

CANYON FORMATIONS

Canyons vary widely in length, width, and depth; are made up of many diverse types of rock and vegetation; and experience different degrees of water flow. All canyons are created from water eroding soil, rock, and vegetation, usually from a streambed cutting through the earth. Other factors that influence canyon topography besides water include tectonic plate movement; soil and rock upheaval; wind, rain, air, and ground temperature; vegetation; animal trails; altitude; earthquakes; landslides; and glacial and volcanic activity.

Preceeding pages: Early morning gear sorting, Kaiparowits Plateau, Utah

High walls are the defining feature of canyons. Some are sloped, allowing easy exit to the rim, while others may be several thousand feet high and sheer, making them all but impassable. A large canyon is usually referred to as a gorge; small canyons or channels cut by running water are, in ascending order of size, gullies, gulches, or ravines. An abyss, or chasm, is a dark, deep hole; this is often an opening to a larger canyon or cave. Cracks, or fissures, are small narrow openings in canyon walls. A couloir is a narrow, steep chute. Ledges, or benches, are the flat features of canyon walls or rims.

Canyons often have brooks or creeks running through them. The swift, turbulent, often seasonal water in a canyon is sometimes referred to as a torrent. An arroyo is a mostly dry canyon cut by occasional water flow.

A slot canyon refers to one that is narrow, sometimes just wide enough for a person to squeeze through; this type of canyon sometimes has water flowing from one wall to the other. Box canyons refer to those that dead-end with no easy escape route.

ENVIRONMENTAL IMPACT

Canyons are usually complex, fragile riparian or desert ecosystems, and it is essential that travelers keep their impact on canyon environments to a minimum. Everyone should do his or her part to preserve and protect these natural formations for future generations and for the sake of animals and plants that make their homes here. When canyoneering, always tread lightly and use good judgment to avoid damage to delicate, wild areas.

Permits now required in many areas

What follows are some general guidelines for low-impact canyoneering. Keep in mind the important adages used by wilderness enthusiasts: "Take nothing but pictures, leave nothing but footprints, kill nothing but time"; "Pack it in, pack it out." Stay informed on issues regarding access or environmental impact by reading and attending fundraisers, lectures, and meetings about canyons and wilderness areas in general. Get involved in environmental causes, and support organizations that promote stewardship of the land. Consider getting involved with a local or national group that protects the canyon wilderness.

Planning

Always use good judgment when selecting a route to minimize impact. Beginners should stick to straightforward routes to prevent damage to fragile canyons. If you are in a canyon that is too difficult for your skill level, it is easy to damage the vegetation or habitat as you try to make your progress easier. Always travel in small groups, since large parties tend to create greater damage to the land, especially when camping. The season should be taken into account as well when selecting a trip; some areas may be better for travel during certain times of year. Always check with managing agencies for regulations and suggestions.

Access

Access to canyons is a difficult issue: you must travel to and from the trailhead, yet cars, bikes, and watercraft can cause damage to the ecosystem. Tread lightly on the approach and use common sense to minimize damage.

ROADS. Roads can damage the ecosystem by altering drainage systems, destroying habitat, and bringing people and cars into pristine areas. Always stay on designated, maintained roads. Don't drive in streambeds or over fragile meadows. Avoid driving on muddy or snowy roads. If you need to travel on an abandoned road that is closed to motor vehicles, plan extra time to hike the distance, or consider bringing a mountain bike as an extension of your car. Always park in designated areas. Plan loop hikes to avoid car shuttles that cost time and fuel.

BIKES. Ride bikes on roads or trails designated for cyclists. Most wilderness areas, national parks, and national forests have rules for bikes. In some areas, including national parks, bikes must stay on roads. Keep in mind that bikes can be hard on trails, macrobiotic soil, moss, and forest duff. Damage from bikes is worse during certain times of year, for example when tires make deep ruts in wet and muddy trails or scrape off topsoil or forest duff. Keep bike groups small, ride single file, and yield to hikers. Stay in control to avoid skids.

WATERCRAFT. Rafters, canoers, and kayakers should make sure their gear is lashed and stowed correctly. Avoid landing on fragile stream banks; attempt a sandy or rocky landing instead. Know the high and low water zones and the flash flood potential. Prevent your gear and trash from floating away.

Closed areas for habitat preservation

Hiking

Employing proper hiking technique is important for low-impact travel, especially in fragile, riparian ecosystems. Try to follow these general guidelines when on trails and around animals and fragile plants.

TRAILS. One of the greatest impacts to a wilderness ecosystem is made by humans hiking on trails. Try to hike in single file and stay on designated trails. Keep your group small to lessen impact. Avoid switchbacks and shortcuts if they are not part of the trail system, as creating new routes promotes erosion. Avoid hiking around a mud puddle on the trail, as this damages fragile vegetation and increases the trail width. If it is necessary to hike off-trail, try to stay on river rock, slickrock, sand, gravel, or snow as much as possible. Steep slopes and wet trails are also prone to damage and erosion. Hiking in water may be less disruptive than doing so on fragile stream banks.

When hiking on a maintained trail, follow the trail markers. These lessen environmental impact by keeping it contained to small areas and helping people from getting lost. Trail markers come in a variety of forms. They may be small signs or markers nailed to posts or ribbons tied to trees. In some areas, small rock stacks called cairns mark the routes. If you need to mark your own route, use natural materials such as cairns or downed tree limbs. Make sure that at the end of your trip you remove all trail markers you put up, and never remove established markers.

FLORA AND FAUNA. Plants and animals are a part of all canyon ecosystems. Sometimes the environment in a canyon can be delicate, and the balance easily upset. Strive to keep plants, animals, and the general ecosystem as undisturbed as possible. Don't pick flowers or move rocks. Avoid fragile ground cover such as moss. Do not tread on macrobiotic soil, that collection of algae, lichen, and moss that covers the desert sand; this delicate living soil is a seedbed for plants, holds moisture, and regulates nitrogen.

With some exceptions, most animals only attack people when in self-defense, for example when their homes are approached, when protecting their young, or when gathering food. Always observe animals from a distance and give them a wide berth.

RUINS. Some canyons, especially those in the southwestern United States, are replete with archeological ruins. Be careful when traveling among these delicate and dilapidated historic structures. Inspect them, but don't climb in or touch them.

Avoid making tableaus, or collections of artifacts such as pottery shards and arrowheads on rocks; it's important to leave relics where they rest. If you do pick up a pottery shard or other ancient object, put it back on the ground where you found it. This is your ethical and legal duty.

View but don't touch petroglyphs and pictographs, the painted or carved symbols and pictures on rock. Hand oil and dirt contribute to decay of these prehistorical treasures.

DEATH HEAT

Desert heat. It was 105 degrees and the sun was scorching my skin, burning my insides. In the shade it was cooler, but not by much. At night the temperature fell a little, maybe to 85 or 90. It was even hotter in my tent than it was outside, but being in the heat was preferable to having bugs bite my flesh, circle around my head, buzz in my ears.

This was my first escapade into canyon country and into a hot, dry box canyon—a death canyon. The guidebook I had with me said someone had died in here from heat exposure: a rancher looking for stray cattle. People die every year because of the sun—from dehydration, heat exhaustion, heat stroke. Once, years later on another trip, I had water intoxication. I felt like I was car sick; I was dizzy and nauseated, all the while guzzling water and urinating constantly. Later, I found out it was 103 degrees in the shade.

On this first trip to the desert canyons, I had no idea the sun could make everything so alive. A whiptail, or maybe a lizard, basked on a hot rock, then scurried to the next rock to soak in more sun. Lichen in psychedelic orange and neon green coated sandstone boulders. Barrel cacti bloomed in pinks and purples. Desert sage put forth a sweet scent, and occasionally some desert lavender or primrose highlighted the dry dirt. Juniper and pinyon trees were tucked in dry cutbacks, their roots going down, deep—very deep—to water.

After this long day, I had sore feet and my pack had rubbed welts on my bare back. I had few desert canyon hiking skills: I brought no long-sleeved shirt and no sun hat, and had no sense to rest during the heat of the day. My two water bottles didn't last half the day. Sunscreen was all that protected my skin from ultraviolet radiation, and it seemed to have evaporated off my skin with my sweat. At a solitary water pocket, I dunked my head and soaked my bandana with stagnant, buggy water. It was more than soothing and refreshing, it was revitalizing and reassuring; I could drink the dirty water if I had to, in order to survive.

Later at camp, the warm water in my car was like a savior. I spent an hour sitting on a sandstone chockstone, long since fallen from the rim, to drink what seemed like a gallon of liquid life.

Waste Management

Personal waste is a huge problem in some areas, especially in the desert canyons where decomposition takes a long time or in slot canyons where one is unable to hide waste from animals and people. In some areas like the Grand Canyon, people have

Climbing the slickrock stairway, Dirty Devil River, Utah

been required for some years now to pack out all human waste, including feces.

If you don't pack out your stool, bury it at least 200 feet away from water in an 8-inch-deep hole. Pack out toilet paper in water-tight plastic containers. When possible, use facilities in parking areas or at ranger stations.

For bathing, carry water 200 feet from the source and always use a very small amount of biodegradable camp soap. Consider using an alcohol gel cleanser or wet wipes to avoid bathing with large amounts of water.

Camping

Campers, especially those in large groups, can cause significant long-term damage to local ecosystems. Make camp 200 feet away from water. Set up tents on sand, flat rock, or forest duff that has no plants if possible. Many areas have specific campsites that are maintained by an agency. Using designated sites lessens impact, since this concentrates use in heavily visited areas.

Try to find a level, sheltered campsite. Avoid using rocks for shelter or tree bows for padding unless you are in an emergency situation. If you must move logs or rocks, replace them when you leave. Store your food in trees away from animals.

Cooking water and scraps should be disposed of properly. Bury your dishwater and scraps in a small hole or scatter it across a large area. Don't wash dishes directly in streams. Collect water in a pan and carry it 200 feet away from the water source. Use a biodegradable soap sparingly.

Fires

Fires use wood that would otherwise decompose naturally, returning important nutrients to the soil. If you must cook or boil water, use a camp stove. Save fires for emergencies, such as when you need to dry clothing, warm up, or cook when you run out of fuel.

Unsightly fire ring

If you build a fire, use only small downed sticks and keep your fire small. Use a dry streambed or dirt hole for your fire. Douse it well with water and scatter soggy ashes around the area. Always dismantle your fire ring or remove your pit.

Pets

Pets are allowed only in some areas. In others only pack animals like horses and llamas may be permitted. Make sure you know the local rules and regulations. Carry and use a leash when appropriate. In general, try to avoid taking your pets into the wilderness, as they tend to contaminate water supplies and disturb the local ecosystem, frightening other animals and trampling fragile plants, soil, and streambeds.

Technical Canyoneering

Low-impact technical canyoneering is not always straightforward. It takes a skilled canyoneer to descend or ascend a technical canyon with minimal impact.

Use natural belays—such as large rocks, logs, and trees—as much as possible. Learn how to use a rock bollard (a stack of rocks), downed logs, or natural chockstones for anchors instead of bolts. Use a retrieval cord to remove your webbing and rappel rings so as not to leave any sign of human passing. If you must leave webbing, try to use tan, brown, or green webbing so it blends in with plants, trees, and rocks. Carry out all old webbing left from previous descents.

As with rock climbing, in canyoneering there is controversy surrounding the use of fixed anchors such as bolts and pitons. The quandary is that some canyons can't be ascended or descended without fixed anchors, but these aids leave permanent destructive scars to the canyon. Some people believe that no permanent alterations should be made in the wilderness even if it means a route could not otherwise be completed. Others believe the impact of a bolt is relatively small when compared to the energy we consume and the byproducts we create when driving cars, flying in airplanes, building homes, making canyoneering gear, and the like. In some wilderness areas, national parks, and national forests it may be illegal to place a fixed anchor or even leave behind webbing and rappel rings. When in doubt, find out the rules and regulations ahead of time. Consider all other options before placing a fixed piece.

SAFETY AND RESPONSIBILITY

No outdoor instruction book is complete without a discussion of safety and responsibility. Mountaineers, climbers, backpackers, and cavers all know they should exercise responsibility and self-reliance. Canyoneers should do so also. Travel in the wilderness is never free of hazards. The idea is to be personally comfortable with an acceptable level of risk and to minimize potential problems.

Risks may be thought of as subjective and objective. Subjective risks can be lessened by choosing safe routes, traveling during clear weather, and using appropriate equipment. Objective risks are those over which humans have less control; there are certain inherent risks in being in the wilderness when conditions are not always predictable. There may be, for example, a landslide, earthquake, or sudden storm that could not have been foreseen.

Strive to be self-reliant and responsible for your own safety. Keep in mind that rescues not only have financial costs—usually on public safety resources—but often have significant detrimental effects to local environments. In some cases canyons have been closed due to the impact of canyoneering rescues. Also remember that rescues are never immediate; there may be delays due to weather, personnel, equipment, or other factors.

Always stay within your skill level and knowledge base. This usually means beginning canyoneering with entry-level nontechnical canyon hikes, and following these with more difficult trips as your skills and knowledge improve. Always keep in excellent physical and mental health. Use good equipment in superior working condition. Be mindful of weather and canyon conditions. Change your plans or cancel your trip if conditions are not safe.

This is not a complete text on canyoneering but an introductory book intended to provide general information. No text can describe every risk or anticipate personal limitations of readers. Nothing substitutes for formal instruction, regular practice, and years of experience. When a reader uses information provided in this book, he or she assumes responsibility for personal safety. This book is a general guide to equipment, skills, safety issues, and techniques in canyoneering. Appendix 2 lists additional resources, including books, magazines, Internet sites, and local sources that provide more detail about the subjects covered in this book. Supplement your learning accordingly with courses on wilderness travel, first aid, mountaineering, rock climbing, and canyoneering. Learning the wonderful skills of canyoneering is an ongoing process that will outlast these pages.

2

PLANNING A TRIP

The first step in—and much of the fun of—a canyoneering adventure is planning the trip. Planning a route takes a great deal of time, especially if scouting is involved. It is never too early to start. This chapter covers the basic procedures for planning an outing. Subsequent chapters detail equipment, navigation, weather, and safety—all important topics to read about and review before you start planning a trip.

An important concept when planning a canyoneering route, and one which will be highlighted several times in this book, is being able to get out of a canyon in cases of emergency or unexpected obstacles. When traveling up or down canyon and you need to get out immediately, you have three choices. You can reverse your route and head back the way you came; you can continue your scheduled route traveling back to your vehicle; or you can hike up to the rim, exit the canyon, and find a new route back. It is important to note also that you may have only one option for exit; your way down canyon may be blocked by a landslide or the canyon walls may be too steep to climb out to the rim. In such a case, heading back may be the only way. Make sure you plan your trip accordingly.

CHOOSING A ROUTE

Looking for a canyon to ascend or descend can be complex and labor-intensive. Expect to spend some time using the resources described below to put together a complete canyon country itinerary.

Guidebooks

The first thing you need to do, of course, is to choose a canyon to hike. Route selection can be made easier with a guidebook. There are several guidebooks written specifically for canyoneering, some covering the southwestern United States and France in particular. However, even general hiking guides, which are more easily accessible, often include canyon routes. If you are heading to an area that has no specific

Columbia River Gorge, Washington

canyoneering guidebook, start with a standard trail guide and supplement with information from other sources discussed below.

When reading general trail books not specifically for canyoneering, start with hiking trails that follow a stream bank or canyon rim. Observe hiking time estimates, canyon entrance and exit points, and information on water sources, major landmarks like waterfalls or arches, and campsites. Keep in mind that routes change because of landslides, floods, and other factors from year to year; some guidebooks may not be completely up to date. Make modifications in the route based on information you gather from various sources.

Maps

Maps are essential for route planning. Good maps are available for almost any spot on earth. Start with maps that cover a broad area, such as those from the National Park Service, National Forest Service, or Bureau of Land Management. These give an overview of an area as well as details of trails or logging and mining roads that can be useful for an approach or exit, especially in an emergency. It is useful to have your entire route on one map so you get an idea of the lay of the land and the area surrounding your route.

More detailed trail maps or those that cover wilderness or other specific areas are usually the most useful, as they show details and are more likely to be updated frequently. Hand-drawn canyoneering maps may be available, although they are generally difficult to find. Usually they are tucked away at ranger stations, outdoor shops, and guide services.

When you get to detailed route planning you will need United States Geographic Survey (USGS) topographic maps (topos). The 7.5-minute series, or 1:24,000 scale, is the most detailed and the standard for routefinding and navigation in the field. Map and compass use are described in Chapter 4.

When planning your route, don't be afraid to mark up your map. You will need to mark both an entrance point and exit point to the canyon, water sources, campsites, emergency exits, and the route to hike back to your car. Roads and hiking trails are often on topos, but they are not always updated. If you want to take one map, write in all changes or additions to your 7.5-minute topo.

Major obstacles such as waterfalls and lakes, as well as significant landmarks such as natural bridges, are sometimes delineated on maps. However, it is important to realize that many obstacles that can block your route are not shown. Seasonal landslides or log jams, small waterfalls or dryfalls, thick brush, narrows, plunge pools, or boulders that may not be on topographic maps can halt your progress. Even a short 10-foot wall or deep plunge pool can make a huge difference in your travel time and equipment needed.

Local sources

Local sources can be invaluable, providing direct information critical to updating your maps and guidebooks. Rangers are perhaps most helpful. Even if they have not hiked directly through the canyon, they are probably familiar with the general area, the rim trails, and seasonal variations. Locals in outdoor or climbing shops may offer advice or direction. Professional guide companies and outfitters are also good resources.

Difficulty and Skill Level

Canyoneering is such a new sport that difficulty levels are hard to determine. Appendix 1 offers a general guide for basic, advanced, and technical canyoneering. This is to help readers get a general idea of types of canyons, their degree of difficulty, and the corresponding skills and equipment necessary to travel in them. The appendix also includes the American Canyoneering Association's rating system, which was recently adopted. Many guide services have their own rating systems, as does the Commission Européen de Canyon. It is useful to be familiar with climbing, mountaineering, and canyoneering rating systems, as you may encounter them in guidebooks.

There are several rating systems around the world for rock, ice, and alpine climbing. In North America, the five classes of the Yosemite Decimal System cover general climbing. Class 1 is hiking. Class 2 is scrambling. Class 3 involves some climbing and scrambling. Class 4 often includes the use of a rope, as a fall risk is possible. Class 5 involves technical climbing and is scaled from 5.0 (least difficult) to 5.14 (most difficult); routes beyond 5.7 require specific rock-climbing techniques.

Alpine routes use the National Climbing Classification System that gives a general route grade from I to VI based on time required, technical difficulty, and risk. A Grade I climb takes a few hours, a Grade II half a day, and a Grade III a full day. A Grade IV climb lasts one long day, and at least one section of the route is a 5.7 technical climb. A Grade V takes more than one day, and at least one section of the route is a 5.8 technical climb. A Grade VI climb spans two days and involves advanced technical climbing.

In general, a canyoneer should be prepared for the most difficult section of a route and allow a wide margin for safety. For example, even if you think that your trek is only a walk-through in a dry streambed, bring along dry socks in case you encounter water. If you know you will have some scrambling and ankle-deep wading but no swimming, down climbing, or rappelling, pack your gear in drybags anyway and carry some webbing or accessory cord to use as a hand line for stream crossings or scrambling.

Know your personal limits with regards to skill as well as those of your companions. You want to plan a trip that is challenging, fun, *and* safe. If you are just learning the sport, pick a descent that you know is free from major obstacles. Find a canyon that has a hiking trail near the bank or which has many exit points to the rim.

In addition to being familiar with the skill level of people in your group, know that you can get along with them and trust their judgment and performance in an emergency. It's best to go for a nontechnical day hike or a practice session in familiar, easy terrain—what is called a shakedown. This allows you to learn the different comfort levels of members of your party so you can supplement or improve skills before a trip. Keep in mind that if you have an accident, you are relying on your companions to help you. People who are level-headed, able to improvise, and have a lot of outdoor experience make ideal canyoneering companions.

Make sure you have the skills to get out of the canyon using one of the three ways described earlier: by reversing your route, continuing up or down the canyon, or exiting to the rim. Remember that if the rim exit is too difficult for your skills, so that for example it requires rock climbing, and if your way down the canyon is blocked by a log jam, you may need to backtrack.

Time

Plan enough time to enjoy your trip and not be rushed though beautiful canyons. If possible, avoid hiking out at dark. Plan to get out of the canyon a few hours before dark so if an accident occurs or the hike takes longer than expected, you have a buffer.

On a dirt trail most people hike 2 to 4 miles an hour. Canyoneering may be slower, especially with obstacles and uneven terrain. Unless you are hiking on a trail or a flat, dry streambed, your traveling time will likely be twice as long as for a regular hike. On some routes, parties travel a quarter of a mile in a whole day of canyoneering. Every time you set up a rappel, swim, or down climb, your hiking time will be slowed. With experience you will be able to estimate your travel time based on canyon terrain, number of rappels and obstacles, mileage, elevation change, and stream flow. Keep in mind that the skill level and physical condition of everyone in your group, party size, and other factors will affect your hiking time.

ACCESS

It is important to make sure before you go on your trip that the road and approach to the canyon you plan on visiting are passable. Also remember to minimize impact to delicate canyon country on your approach. Check the rules and regulations for visiting well in advance of your trip. This information is available by telephone or via the Internet for the various national parks, national forests, and wilderness areas. The websites for major land management agencies in the United States are listed in Appendix 2. Also, always get permission if you plan to travel through private land.

You probably will need to register at a ranger station or trailhead or get a permit (with or without paying a use fee). Some areas require a parking or camping permit only; others require a day use fee. For camping in popular areas you may need reservations.

CURRENT CONDITIONS

Conditions are widely variable and constantly changing. Having a safe canyon adventure depends on many factors, including the nature of the canyon, how accessible it is, and perhaps most importantly, the weather.

Canyon Conditions

Check physical conditions of the canyon, rim, and approach well ahead of time. Watch the seasonal snowpack, weather, and other factors that influence water flow down

Wild River, Columbia Gorge, Oregon

the canyon stream. If the canyon is fed by a reservoir, call the water district to find out this information. If the canyon is fed by glaciers or mountain drainage, check the status of the runoff and its effect on water level and flow. Some canyons may be impassable or dangerous with high water flow. Check for seasonal alterations such as recent landslides, flood damage, chockstones, or log jams.

Weather

It is imperative that canyoneers take the weather seriously. It is important to learn the typical weather patterns of the area you will be visiting; you should know, for example, if a thunderstorm miles away can cause a flash flood in your vicinity. Watch the weather months in advance to find out about seasonal changes. Always know the weather forecast for the days you will be canyoneering. Try to get a last-minute update, as weather can change suddenly and unpredictably.

The Internet can give you up-to-date weather patterns typical of local areas; see Appendix 2 for a list of websites to visit. Because knowledge of the weather is so vital to canyoneering, Chapter 5 is devoted to basic weather reading and forecasting techniques.

Roads

Be familiar with road conditions in the area you will be visiting, and make sure they are open. Learn the conditions of unimproved roads and the effect the weather may have on them. An evening thunderstorm can make a road on clay-based soil impassable even with four-wheel-drive vehicles. At higher elevations snow can fall when you least expect it and in all seasons of the year.

SCOUTING

When planning a trip other than one described in a guidebook and marked on a map, you will probably need to do some scouting. The best time to scout a canyon is the same season and year you plan to hike it. Plan an extra day or two at the start of your trip to check out canyon conditions, or make a separate scouting trip weeks or months in advance. However, keep in mind that seasonal floods, runoff, landslides, and other natural phenomena change canyon terrain. What may be passable in the spring may not be in the summer, and vice versa.

Make sure you scout exit points where you can hike out to the canyon rim. You may have to walk the entire rim and hike down into the canyon floor to make sure that an exit point is passable. A mere 10 feet of rock wall can block an exit, and it's best to know this ahead of time. In extreme cases, you may need to hike from the rim down to each rappel to determine length and difficulty.

WATER

Water is a precious resource in canyon country. Planning water requirements and knowing about available sources is vital to a safe trip. Your daily requirements depend on many variables, including exertion level, outside temperature, elevation, and your personal metabolism. A rough estimate for water requirements per person is a gallon daily. However, in hot deserts, cold environments, high elevations, and on trips with intense physical exertion, you may require more. On straightforward canyon hikes, you may require less.

Before you head out, make sure you are well hydrated. This means on the drive up as well as on the days before you go. The amount of water you plan to carry in your pack will depend on your destination. Always have enough to get you safely to a known water source. Confirm the location of that source from rangers or locals who have made the trip recently. If the canyon you will be visiting has running water, this may mean you only need to carry a quart or two per person and equipment to filter or purify more. If the canyon has no running water, make sure you have enough to get to the next spring, pool, stock tank, or water pocket on your route. If the canyon is potentially dry with no

Hole in the rock

reliable water source, carry water for the entire hike. On extreme trips, you may be able to place water caches strategically along the route. Make sure you have a supply in your car for the end of the trip. Never pass a water source without filling your bottle in desert canyons.

Some athletes use an electrolyte drink such as a commercial sports drink. With an electrolyte solution, you can absorb water efficiently and replace salt and other electrolytes that you lose when you sweat. This is important when treating dehydration and heat exhaustion, as noted in Chapter 6, or when hiking in a hot desert environment, where you lose electrolytes through sweat.

Procuring Water

If you collect water on a canyon trip, you should purify or filter it. This can be done by boiling or using chemicals, a filter, or a purifier. Whatever the case, plan ahead.

Boiling water kills most microorganisms. However, you will need a stove and fuel, which you may not want to carry on a day trip.

Chemical purification is a simple alternative that is light in weight and inexpensive. Iodine tablets, which are available at camping and mountaineering stores, kill germs within an hour or two of being dissolved in water. However, iodine-laced water tastes terrible, and too much of this chemical may be detrimental to one's health.

Filters and purifiers are readily available in outdoor recreation stores. These compact devices do add some extra weight, but they are not as bulky as stoves. A filter eliminates many germs, including bacteria and giardia, a microscopic parasite common in the mountains. Filters are usually made of ceramic or a composite material and require routine cleaning and maintenance. The ceramic filters are a bit heavy but will likely last a lifetime if not cracked from freezing temperatures or from being dropped on a rock. Cartridges in the composite filters must be replaced regularly.

Filters don't purify water, because viruses that are too small to be trapped escape through the small pores. In some areas, especially in tropical and subtropical regions, complete purification is essential. A purifier first purifies with a chemical-impregnated filter. These systems eliminate virtually all germs and some heavy metals. Like filters, these combination units must be cleaned often and the cartridges replaced regularly.

FOOD

The outdoor athlete's diet is the subject of many books, and there is heated debate about what types of food are most beneficial. Carbohydrates include bread, grains, pasta, fruits, and vegetables. These are easy to digest, and simple carbohydrates in the form of fruit or candy provide quick energy. Protein-rich food such as dairy products, nuts, and meat, which are necessary to build muscles, take much longer to digest. Food high in fat gives more calories with less weight; one gram of fat has twice the calories of a gram of protein or carbohydrate. Thus high-fat foods such as cheese, meat, and nuts are loaded with calories.

The best solution is to plan a diet that contains high-quality, high-calorie food that is to your liking. Lightweight camp foods are easy to prepare. If you choose not to take a stove or if you are on a day trip, keep it simple. Pre-made sandwiches and trail mix, dried fruit, nuts, or other snacks make an easy lunch. Remember to eat well on the days before you head out on a trip. Dinner the night before and breakfast the morning you leave are especially important. Try to plan every meal as well as snacks. Also, bring emergency food for an unexpected night out.

ALIENS IN THE PARK

We must have looked as if we were from *The X-Files*. It was an August morning and we were at a national park entrance packed with recreation vehicles and sightseers hiking the quarter-mile trail to a scenic view point. We squeezed into full-length neoprene wet suits and donned our day packs complete with ropes, webbing, and climbing hardware that we'd spent the morning organizing. Hiking past the ranger booth and down the trail, people moved out of our way, despite our leisurely pace. They could hear us coming, half a dozen of us jingling with rappelling gear, dressed in wet suits in the 90-degree desert sun. They looked at us quizzically. What on earth were we doing? Near the viewpoint we wiggled into our harnesses, slung the tree, and began the drop into the dark canyon.

The first drop is always the scariest. After pulling the rope, there's no turning back. No way to get out but down—unless your way is blocked by boulders, landslides, log jams, water, or a dead-end blank wall. My turn to rappel came first and, despite years of climbing experience, I had a shudder of fear. Down into the dark slot claustrophobia attacked my fear of heights and struggled for a foothold in my head. Adrenaline surged. Down, down, darker, darker. I hit water: murky, cold, and black. The plunge pool was filled with logs, dirt, and other debris. The air was so stagnant it made me squint; I tried to keep my breathing as shallow as possible. I didn't want the water touching my bare skin. I unhooked from the rappel while floating and swam for a small ledge of rock. Before I knew it, my head got dunked and my hair was soaked with nasty slime. On a ledge of rock and sand, I waited for everyone else to rappel, too pumped on adrenaline to remember to pull out my camera. When we pulled the rope, I watched it spindle in the air for a microsecond as if to say, "You idiots. You alien idiots."

We headed down canyon following a poor photocopy of a hand-sketched topo map likely made after the first descent years ago. Four rappels and a few swims later we came out below the viewpoint: soggy, wet, tired. Hands shredded, equipment trashed. We followed the stream in the hot sun and stripped at the first pool to wash off the slime and heat. Not until my eyes spied the white truck did I even begin to come off my adrenaline buzz.

BEFORE YOU GO

Any wilderness trip can be fairly complicated. You definitely don't want to forget any vital equipment or miss talking with the ranger about canyon conditions. Consider making a to-do list before each trip. Here is a list of things to cover so you won't forget anything.

Deep, going deeper, Robbers Roost, Utah

- Check your gear and replace all worn or damaged parts. Do this when you return from a trip so your gear is ready for the next outing. Don't forget to replace batteries in your headlamp, cell phone, GPS, or camera.
- Confirm campsite reservations or permit availability.

- Check and re-check weather, canyon, and road conditions, noting especially any seasonal changes.

- Leave your route information with someone. This should include your route description, possible changes due to emergency or impassable sections, your expected time of return, and potential problems.

- Make sure everyone is healthy. Nobody in your party should be sick or not 100 percent recovered from an injury.

- Never hesitate to get more information. After all the planning, packing, scouting, and organization, you still may need more information. Talk with rangers again, review maps and guidebooks one more time, or plan a day to scout entrances or exits to a canyon. If you meet people in the parking lot, try to share information.

- Cancel or postpone the trip if necessary. Don't be so locked into a time and date that you are inflexible. Many people get lost or injured because they set out when they should have stayed home. Don't be afraid to postpone the trip if conditions are not safe, someone in your group is ill, or for some other reason.

3

EQUIPMENT

Assembling canyoneering equipment is part of the fun of the trip. With so many out-door gear supply companies, it can be confusing, too. A few manufacturers design equipment specifically for canyoneering, such as harnesses, shoes, drybags, and back-packs. Since the sport is relatively new, much of the equipment is adapted from rock climbing, caving, mountaineering, and backpacking.

When shopping, keep in mind some general guidelines. Look for well-made equip-ment manufactured by a reputable company and check around for good deals. Your local mountaineering or camping store is a good place to start. Employees at the shops can help you fit items and discuss options. Some mail-order companies carry hard-to-find specialty equipment.

Buying equipment sometimes requires a trade-off. Oftentimes, the lighter or more du-rable the gear, the more expensive. However, lightweight gear sometimes sacrifices durability. High-performance gear that is designed for a specific application can be costly as well. You may be able to find equipment that has multiple uses, for example a com-pass that is also a clinometer and signal mirror, or shoes that double as dry- and wet-hiking footwear. However, multitask tools sometimes compromise performance of one use or another.

In general, try to buy the best gear you can afford. The better the quality, the longer it will last. You may be able to rent or borrow some items before you buy them. Some gear you might be able to find used, or you can adapt some equipment you have for backpacking or climbing. Keep in mind that canyoneering is a rigorous sport and takes a significant toll on gear, especially if it is not designed for such use. Try not to borrow, rent, or buy used technical canyoneering gear or other equipment that you will have to depend on for your safety.

It can be tough to decide how much gear to bring. You will need equipment to keep you safe, healthy, warm, and dry. More gear sometimes improves comfort and can give a better margin of safety in case of a mishap. But you may find that you cannot carry comfort items that are not absolutely necessary. A lighter pack can mean quicker hiking time, less fatigue, and sometimes safer travel through canyons. You will need to find both a personal and group balance for how much equipment to bring, especially with regards to comfort items, emergency supplies, and safety gear.

Finally, know how to use all your equipment. Take a course to learn the proper techniques for using all the gear you bring, especially for safety, technical, and first-aid equipment.

ESSENTIALS

Certain supplies you will need to carry every time you head into the wild. These, known as essentials, are listed in Table 1. Most canyoneers have similar basic recommendations, with some modifications depending on mode of travel, climate, and other factors.

Table 1

GENERAL EQUIPMENT FOR CANYONEERING*	
ESSENTIALS	
Food	Sunscreen and lip balm
Water and water bottles	Headlamp with extra batteries
Map, compass, and altimeter	and bulb
Windproof and waterproof	Water purification tablets or filter
matches, lighter, and fire	Extra food, water, and clothing
starter	Emergency bivouac sack or
First-aid kit**	plastic tarp
Repair materials and multi-use	Backpack
tool	Drybags
Sunglasses	Throw rope or hand line
CLOTHING	
Sun hat	Rain gear for wet weather
Fleece hat	Thermal long underwear for
Balaclava, neck gaiter, or	cool weather
headband	Insulating layers of polyester
Bandanna or cravat	fleece for cool weather
Lightweight pants for hot	Gloves
weather or insects	Gaiters
Long-sleeved shirt for hot	Socks, neoprene, and/or wool
weather or insects	blend
Shorts	Footwear, boots, and/or sandals
Underwear	Wet suit or dry suit

TECHNICAL HARDWARE

Hiking Poles

Helmet

Harness

Rappel/belay device, plus a
    spare

Carabiners, locking and regular

Rappel rings

Washers and screw link

Pulley

Rope

Webbing

Cord

Ascenders and aiders

Bolt kit and/or piton kit

Rock-climbing protection

OVERNIGHT GEAR

Stove, fuel, and stove-cleaning
    kit

Cooking pot, pot handle, bowl,
    spoon, and biodegradable
    soap

Water filter

Sleeping bag and pad

Sleeping bag liner or cover

Tent or bivouac sack

Ground cloth or tarp

Additional food

Additional clothing

Personal toiletries

Trowel and toilet paper

VEHICLE ITEMS

Spare tire

Jack and tire iron

Sand plate or carpet

Jumper cables

Tow chain or strap

Repair tools

Spare parts

Extra gas

Extra motor oil

Extra food and water

Tire pump

Shovel

Fire extinguisher

Portable light

Flares

Rope or tie-down straps

ELECTRONICS

Cell phone

Global Positioning System unit

Camera and film

* For bike and watercraft, see Tables 3 (pages 129–130) and 4 (page 136), respectively.

** For first-aid kit recommendations, see Table 2 (page 73).

The "extra" food, water, and clothing on this equipment list is above and beyond what you carry for your scheduled trip. Depending on the length of your trip, the distance from help, the number of people in your party, the technical skill involved, and other factors, your first-aid and repair kits may vary in contents. Chapter 6 and Table 2 (page 73) cover first-aid kits. Bare-minimum repair materials include duct tape, cloth first-aid tape, a rubber strap, a few feet of cord, wire, a safety pin, and a pocket tool. Tables 3 (pages 129–130) and 4 (page 136) cover extra equipment for biking and water travel, respectively.

Keep in mind that you should know how to use every item on your essentials list, especially your first-aid kit, repair supplies, map and compass, fire starting tools, and water purification device.

CLOTHING

Your clothing is some of your most vital equipment in canyoneering. It falls into two general categories: clothing for cold and wet weather, and clothing for hot and dry weather. Canyoneering trips in water may require wet suits. On some trips, for example to the desert in the spring and fall, you will need both sets of clothing: hot days require sun protection and cool nights or thunderstorms may demand warm clothes and rain gear.

Cold- and Wet-Weather Clothing

Synthetic clothing crafted from polyester is now the backcountry standard for insulation from cold. Polyester fleece and its variations wick moisture well, dry quickly, and retain some of their insulating properties when wet. They are lightweight and durable. Down clothing is light and compact, and is an efficient insulator in cold, dry climes. However, when soaked with water, which is often encountered in canyoneering, down becomes a poor insulator. The old outdoor standard, wool, is durable but heavy, especially when wet.

Layering clothing is the most efficient method when trying to balance internal heat and sweat production with outside cold and precipitation. A bottom layer of thin polyester or polypropylene underwear will wick moisture from your skin. A good middle layer consists of thick pile fleece. Sometimes a lightweight fleece vest works well for the middle layer; other times you will need two thick fleece jackets or sweaters. The outer layer is usually a windproof, waterproof rain suit or parka and pants. This final layer should keep you dry from rain or snow but allow for some venting of body heat and moisture. In cold weather, use a hat, hood, headband, balaclava, and gloves for additional protection. Keep in mind you lose a great deal of heat from your head, so a hat is important.

Hot- and Dry-Weather Clothing

Hot and dry weather requires clothing to serve two basic functions: to keep you cool and to protect you from harmful rays of the sun. You should always carry a long-sleeved shirt and lightweight, light-colored pants for hiking in the direct sun. It will protect you from the sun's dangerous ultraviolet rays and keep you cooler by reflecting heat. Also, wear a wide-brimmed hat, or cover your neck and head with a large bandana.

Some people believe that cotton should never be used in the backcountry, as it is a poor insulator and dries slowly. However, the one exception is perhaps desert hiking. Cotton breathes well and is lightweight. And if you soak your cotton T-shirt in water, the evaporation can help keep you cool. An alternative fabric for warm weather is a synthetic or cotton-synthetic blend designed for hot-weather hiking.

FOOTWEAR

Like clothing, footwear is some of your most important canyoneering equipment. For any trip, especially in a survival situation, it is crucial to have proper shoes and warm, durable socks.

Shoes

The type of shoes you choose to wear depends on the difficulty of your route and the nature of the terrain. Several different types of shoes and boots are available. Most are a

Canyoneering shoes

trade-off between durability and support and weight and performance. Your shoes or boots should provide grip and cushioning but should also withstand the extremes of canyon hiking, which might include rock climbing, water hiking, and swimming. Whatever you choose, make sure your footwear fits well, especially in the heel cup and toe box.

Canyoneering or amphibious shoes, which are now available from several companies, are perhaps the best option, especially for technical routes. These are usually wet- and dry-approach shoes or lightweight hiking boots with sticky climbing soles, neoprene uppers for warmth, and extra reinforcement in the seams, toes, and heels. They are designed specifically for the rigors and varied terrain of canyoneering and make it possible to take one shoe for multiple types of terrain. They are great for water hiking or mixed terrain that includes water and slickrock. They are somewhat of a compromise when it comes to long-distance hiking over rough terrain.

Medium- or lightweight canvas and leather hiking boots with rubber soles work well. They provide support and cushioning for difficult hiking with a heavy pack. Although durable, when waterlogged they can be quite heavy and slow to dry.

Approach shoes, which are low-top, lightweight hiking shoes, are a good alternative. They provide a balance between support, durability, and performance. Most have beefy stitching and sticky rubber soles that grip well. Tennis shoes are similar to approach shoes. However, tennis shoes are not as durable or supportive, especially for hiking in water, climbing steep technical terrain, or carrying a heavy pack.

High-tech sports sandals also provide both cushioning and rubber soles for traction. They work moderately well for nontechnical day trips that involve a lot of water hiking on flat ground with a light day pack. Sandals don't provide much support, especially on lengthy hikes, with a heavy pack, or on technical terrain. Also they don't give much protection from thorns, rocks, and snakes.

Socks

Wool-blend hiking socks are the standard for durability and warmth on dry hikes. For hiking in streams, neoprene socks are much warmer. Some companies make waterproof socks with impermeable membranes sandwiched in nylon. These are less durable, however, and a leak can cause them to fill with water. If you are hiking in dense brush or in areas where ticks may be prevalent, consider using gaiters to keep your shoes and socks clean. Fleece socks are warm, and they dry quickly when wet. However, fleece is not as durable for long trips and is too hot for summer hikes.

BACKPACKS

Backpacks are available from numerous companies. A pack should be durable and have heavy-duty stitching. Look for backpacks made with sturdy material and reinforced

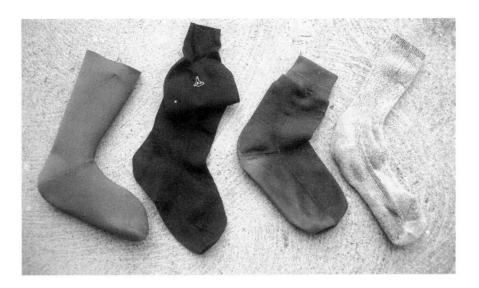

seams. Internal frame packs are more durable than external frame packs, and are espe-cially good when you are sliding down rock, squeezing through slots, hacking though brush, and swimming through streams. Because of the varied terrain you will encounter while canyoneering, get a pack with minimal external accessories. Shovel pockets, water bottle holders, and extra pouches are convenient if you are hiking on flat, dry stream-beds, but they can get torn off when you are squeezing, sliding, or swimming.

Your pack should be custom-adjusted to your body. Backpacking stores usually have experts who can help you with a correct fit. Many packs are widely adjustable and come in several sizes.

When choosing size, consider a pack large enough to fit all your canyoneering gear inside. Anything strapped to the outside can get damaged. The size you choose will depend somewhat on your trip length and the technical gear required. Some people choose ultralight camping and get by with a smaller pack. Others want to be able to pack in all their overnight gear and a week's worth of food. For nontechnical day hikes, a 1,500 to 2,000-cubic-inch pack is large enough for essentials, food, water, camera, and extra clothes. You will need a larger pack, something around 2,000 to 3,000 cubic inches, to fit in technical gear and a wet suit even on day hikes. For over-night trips your pack may range from 3,000 to 7,000 cubic inches, depending on the number of days you will be outdoors and the amount of equipment you require.

DRYBAGS

Drybags are necessary for any trip that involves swimming or significant stream cross-ing. Dry clothes, food, and sleeping bag provide more than comfort; they can be vital to survival. Most stores that sell kayaking or rafting gear carry drybags in a wide variety of sizes and styles. For canyoneering, the thick, heavy-duty drybags are the most suitable. Store your gear in several small drybags; this allows you to access some gear without opening and digging through a large bag. Also, if one leaks, at least some of your gear will stay dry. Another option is buy one big drybag the same size as your pack. A few companies make drybags that have backpack straps, for another option. Always pack essentials like food, maps, and a first-aid kit in thick freezer bags as backup in case a drybag leaks.

For a less expensive, less durable option than a commercially-made drybag, wrap your gear in plastic bags, then put it in stuff sacks. Line your pack with a large garbage bag or two.

WET SUITS

Canyon streams are almost always cool, and pools deep below the rim can be frigid despite hot air temperatures on the plateaus. Even polyester fleece is often not adequate to keep you warm if you plan to spend much time in the water. If in doubt about water temperature or depth, carry a wet suit. Neoprene wet suits designed for windsurfing, surfing, water skiing, or diving are durable yet allow for adequate movement of arms and legs. However, they don't keep you totally dry. Rather, they keep you warm by heating the water next to your skin. For this reason, a small puncture or tear does not effect their heat-retaining capacity much.

The style and thickness of the suit you choose will depend on the amount of time you are going to spend in the water, as well as the water and air temperature. The Farmer John, or long-legged, sleeveless suit, is the least expensive basic wet suit that is appropriate for canyoneering. It can be used with or without a separate jacket in most situations. The Farmer John keeps your legs warm when you are hiking in waist-deep water, but gives you freedom of movement in your arms.

If you are looking for something warmer, you will find a large selection of long-legged, long-sleeved suits ranging in thickness from 3 to 6mm. Suits designed for active sports in cold water such as ocean surfing or windsurfing have thicker neoprene on the torso and thinner neoprene on the arms and legs for mobility. Cold-water diving wet suits are probably the warmest, thickest suits; they are durable but bulky and expensive.

For a trip on which you're not expecting to get wet very much, a 1-mm neoprene shirt or a kayak top may be adequate. This can also add an extra layer of warmth when worn under a wet suit. You can get a shorty, or a short-legged, short-sleeved suit, for a bit more protection.

Dry suits are usually made from waterproof rubberized nylon or a similar fabric, and have gaskets that totally seal out water. They keep you warm and dry until they are punctured, at which point they then fill up with water. These are generally not recommended for canyoneering.

Wetsuits (l to r): Full, short-sleeve, shorty, shirt

FLOTATION

In many situations, streams or pools can be crossed by wading. If you need to swim or float, you can use your backpack filled with drybags for some flotation.

However, in canyons in which you will encounter significant water or large pools, you may choose to bring along additional flotation devices. A life jacket is bulky but probably the safest. Some people bring inner tubes or air mattresses to use for flotation and to keep their gear dry when swimming. Another option is an inflatable dive vest or even an inflatable raft.

POLES

Hiking or trekking poles are of great benefit in canyoneering. They provide extra support and help you conserve energy, especially when you are hiking down steep slopes, on slippery rocks, with a heavy pack, in water, or over uneven terrain. You can also use them to probe pools for depth or submerged rocks or logs.

Collapsible poles can be adjusted in length, which is nice when you are hiking up or down a hill. They are also easily stowed in your pack when not in use. Try to find poles with replaceable tips, baskets, and shafts so that when they break you can replace parts. Old ski poles are an inexpensive alternative.

Use small trekking baskets on your poles or remove them altogether. Large ski-pole baskets can get stuck in holes and among rocks. Use wrist straps, especially with water walking so you don't lose a pole.

OVERNIGHT GEAR

If you plan an overnight trip you will need extra clothes, more food, sleeping gear, a cook set, and a larger pack to store it all in. Here is an overview of supplies you will need.

Cook Sets

Some ultralight-travelling canyoneers prefer not to carry a cook set. A hot meal at the end of the day or a cup of coffee in the morning is not a necessity for everyone. Unless you need your stove to melt snow and boil water, often you can get by without it.

If you elect to cook, there are many types of stoves to choose from. Most run on white gas or isobutane. Some use or accept other types of fuel such as kerosene, propane, or auto gas. In general, 0.3 liters of fuel will last one or two people for one to two days for a hot drink at breakfast and a hot dinner at night. If you are thrifty with fuel it can last longer. If you need to melt and boil snow for drinking water, you will need more. A windscreen can help save fuel as well. In addition to a stove you will need a cooking pot, a pot handle, a bowl, and a spoon. Don't forget to also bring along a stove-cleaning kit, a lighter, and windproof, waterproof matches as backup.

Sleeping Bags

Sleeping bags, like backpacks, are made by a number of manufacturers. They are of varying quality and come with different price tags. With sleeping bags, you have a choice of fill material as well as weight.

Synthetic-filled bags, which are usually made with a type of polyester fiber, are the choice of many canyoneers. Synthetic bags do not absorb nearly as much water as down bags, so they dry more quickly and still retain insulating properties when wet. They are generally less expensive but may not last as long. They don't compress as well and are heavier for the same warmth rating as a down bag.

Down bags are more or less the standard when it comes to mountaineering high-alpine environments. When compared to synthetic bags of equal warmth rating, down bags are lighter and compress smaller. They are more expensive but are more durable and last many years. Unfortunately, when down sleeping bags get wet, they are poor insulators and they take a long time to dry out. Hence a down sleeping bag is not the best choice for a canyon with a great deal of water, unless it is packed in a durable drybag.

Consider a four-season mountaineering bag for extreme temperatures. The chief difference between this and other bags is it has more fill material and is constructed to keep out drafts. A three-season bag works well for most trips. You can always use a sleeping bag liner, cover, or extra clothes for more warmth.

Sleeping Pads

A sleeping pad will not only keep you warm but make your bed more comfortable. Much of your body heat is lost to the ground when you sleep in the wilderness, so a pad is important.

Closed-cell foam pads are cheap and durable. Thicker pads are more comfortable and warmer, but bulky. If you must strap your closed-cell pad to the outside of your pack, it will probably get torn up a bit but it will still function.

Inflatable pads are used by many backpackers. The advantage of these over foam is that they compress, are easier to stow in a pack, and can be more comfortable. Be careful packing an inflatable pad, however, as a hole can render it useless. Store it inside your pack for protection and bring along some repair tape.

Tents

Most people bring tents for overnight trips. Not only does a tent protect you from wind, rain, and snow, it keeps you warmer and provides protection from bugs as well. Fortunately, a multiperson tent can be disassembled into many parts and carried by several people.

Tents, like packs and sleeping bags, come in many shapes and sizes. You will have a difficult time deciding among them. Your basic criteria for choosing a tent are how big and how warm you want it to be. A four-season mountaineering tent is the warmest and most durable. For most canyon trips, a three-season tent is adequate unless you expect winter conditions. How big a tent you need depends on how many people you expect to sleep in it. Most people carry a two- to three-person tent for small groups. With two people, there is usually room for packs and gear inside. Three people can usually squeeze in with gear outside.

Dome tents are the standard when it comes to style and construction. A freestanding tent, or one that does not need guy lines or stakes to be set up, is ideal for canyoneering. Non-freestanding tents, those that need to be staked down, take longer to set up but are often lighter and less expensive. These tents require the added weight of stakes or have guy lines that must be tied to rocks or trees. In some cases, you may not be in an environment where there is soft soil to drive in stakes, or rocks or trees for guy lines may be unavailable.

In general, tents are designed with two layers. The tent itself is permeable to allow body moisture to escape and minimize condensation. The waterproof rain fly keeps you dry. Look for a tent with insect doors, durable zippers, and a thick floor. Vestibules give you more room for cooking or storing your packs. A separate ground cloth provides protection from rocks and sticks. Skylights are nice to give you a view of the stars.

For less strenuous trips, especially those in summer, a tarp, bivouac sack, or one-person tube shelter may be enough. Some "deluxe" bivouac sacks have bug screens and room for personal gear. Also, keep in mind that in some canyons there may not be room on the canyon walls or stream banks for a tent.

TECHNICAL GEAR

Personal rappel gear

Technical canyoneering is a unique and thrilling sport. It usually involves some advanced level of negotiating canyons. Usually this means rappelling, ascending, and rock

climbing. Rappelling, or sliding down a rope, is used to get down a drop too steep or too high to climb down. Ascending is a technique employed to go up a rope. Rock climbing uses specific skills to go up sections too steep to simply hike. Technical gear can be described and discussed in great detail but is only introduced here to give readers an idea of the basic equipment necessary. The use and application of this equipment is discussed in Chapter 9.

Some companies make specific canyoneering equipment, but most hardware is adapted from caving, mountaineering, and rock climbing. Check out a book that covers those sports for more information.

As noted earlier, this gear will protect your life, so always buy new equipment and keep track of its use. Retire and replace items when they show wear. Also, technical gear should be certified by the Union Internationale des Associations d'Alpinisme (UIAA), an organization that tests, rates, and certifies climbing gear.

Harness

A harness, which is made from webbing, wraps around a canyoneer's legs and waist. It is used to provide the means to attach the rope to a canyoneer for rappelling and ascending or to protect against a fall when rock climbing. A harness designed

Simple chest harness

specifically for climbing or rappelling works well, depending on the nature of your route. Find a harness with adjustable leg loops and waist belt. Make sure you find one that fits well, as they come in several sizes. Rappel harnesses often have a carabiner attachment above the belt; these models are more comfortable for trips that only require descent.

A chest harness can be useful for keeping you upright, especially if you wear a pack during rappel or have a long rappel. You can improvise a chest harness by making a figure 8 with a long piece of webbing. Put each arm through a loop with the crossed piece resting between your shoulder blades. Clip the two loops at your chest with a carabiner and then clip the carabiner to the rope.

Helmets

A helmet is necessary during any canyon trip that may include dangerous rappels, rockfall, or the risk of a fall. Those designed for rock climbing or mountain-

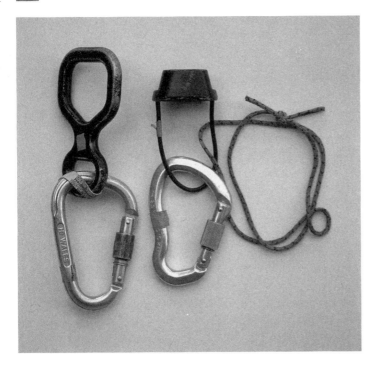

Figure 8 with rubber band, tube descender with leash

eering usually suffice, as they are lightweight, durable, comfortable, and allow for ventilation. Make sure your helmet fits well and is adjustable to allow for a hat in cold weather.

Rappel/Belay Devices

The standard tool for attaching the rope to the harness is a rappel/belay device, which may be either a tube or a figure 8. These metal devices allow the canyoneer to slide down the rope in a controlled fashion, slowing or even stopping the rope in a certain situation like a fall. A tube is easy to use and durable. Figure 8s are favored by some because they allow for several different breaking techniques as described in Chapter 9.

It is easy to drop your tube or 8, for example when unclipping from a rappel while floating in a pool. Lash it to your harness or carabiner with a cord or rubber band. Always carry a backup tube or 8 or learn how to improvise using a carabiner as a rappel/belay device.

Carabiners

Carabiners, the large snap links used in climbing, come in many varieties. Carabiners are necessary for clipping gear, setting up rappels, lowering packs, and performing rappels and belays. In general you will need at least one locking carabiner (the opening of the carabiner or gate screws shut) for attaching a rappel/belay device to your harness. Locking carabiners are stronger than those that do not lock and provide an added measure of safety against opening; this is essential for any application in which you need the extra precaution against your carabiner opening.

Regular carabiners, those that don't lock, are either D-shaped or oval. D-shaped carabiners are a bit lighter, but ovals hold gear with less shifting. Two oval carabiners, side by side, with the opening latches, or gates, reversed and opposed can function in place of a locking one.

Ascenders

Ascenders are devices used to climb up a rope, for example when a climber is backtracking after a rappel or following another climber who has already placed a rope. Mechanical metal ascenders come in several varieties; most clip onto a rope with a spring-loaded clamp. An alternative to using an ascender is to use two pieces of accessory cord or webbing tied in a prusik or Klemheist knot, as described in Chapter 9.

Other Hardware

In addition to a rappel-belay device, carabiners, and ascenders, a few other items are required for technical canyoneering. Rappel rings made of aluminum or steel are used for setting up a rappel station. These allow the rope to slide easily, minimizing friction and damage that may occur if the rope runs through webbing alone.

In some situations, a screw link, or lock link, can be used as an alternative to a carabiner. This is smaller than a carabiner. The gates of the link screw closed for security. Washers may be used if you are using a retrieval cord to pull a rope on a single line rappel, as described in Chapter 9. Pulleys come in handy for lowering or raising packs or setting up rescues.

Rappel/belay devices (l to r): Tube, figure 8, carabiner wrap, Münter hitch

LITTLE PEOPLE

Traces of little people were scattered down the canyon. Their spirit drifted in the air. It was an aura that kept my wife and me company and gave us a mission for the trip. After we started hiking at our usual fast, steady pace, we found ourselves slowing and stopping frequently for ruins or rock art. Ancient pueblos constructed mostly from mud and sandstone were in various states of preservation.

Sometimes we'd find a full house with kiva and cooking area, other times we would stumble upon a spot with only traces of structures. Rock steps, called *moqui* steps, might lead to a higher perch such as a granary or lookout; other times we could not tell where the steps led.

Outdoor kitchens were marked by the presence of a boulder with a smooth trough from grinding; sometimes nearby we found an intact grinding stone. Occasionally all we saw were fragments of the life that was: miniature corncobs, arrowheads, pottery shards. In some cases, past canyoneers had collected and arranged them on a rock, a tableau.

The word Anazasi means, according to some translations, the ancient ones; others say it means the enemy. We called the folks that lived here the little people. Their abodes had low roofs—not tall enough for us to stand in—and miniature doorways so small that, if a present-day visitor climbed in (which she is not supposed to do), she would have to squeeze in and crawl.

Most interesting perhaps were the pictographs and petroglyphs, drawn or carved on rocks in the canyons. Overgrowth of poplars and cottonwoods as well as the varnishing of rock panels made them difficult to find. We tried to decipher whether they were for communication, ceremony, history, or art. We occasionally found markings from early American explorers; they had carved their names and dates

Rope

Rope comes in a variety of strengths and lengths. Modern rappelling or climbing rope has what is called kernmantle construction: it has core, the kern, and outer sheath, the mantle.

Two types of rope are used for canyoneering: static and dynamic. Static or low-stretch ropes with about 3 percent elongation are the standard for canyon descents that require rappels only. These are not designed for lead climbing. Dynamic, or elastic, ropes, which stretch significantly, are designed for lead falls. These ropes must be able to stretch enough to gently slow a climber's fall but must be durable enough to stop him or her quickly. Dynamic ropes are more expensive and heavier than static ropes; they have about 6 percent elongation.

in the rock, marking the sites as their own—even though they belong to no one. Present-day scrawls and bullet holes vandalized some roadside rock art.

Once, inspired perhaps by the rock art, I etched my wife's face in a sandstone boulder using a small, hard rock. I carved a portrait of my lover in a warm, sunny resting spot using tools of the earth. Oblivious to canyon ethics, I was, I suppose, trying to get closer to those people that once lived here, to recreate some ancient aura, to make a bond to their mysterious world.

Ancient rock art

Static ropes are generally available in thicknesses of 9 to 12.7 mm for rappels. Dynamic climbing ropes range from 9.5 mm to 11 mm in thickness. Some ropes, such as "double" or "half" ropes, are designed for special climbing techniques.

Some ropes are available with a dry finish that decreases absorption of water. This is important as a rope can lose up to 30 percent of its strength when wet.

Webbing and Cord

Webbing is essential for setting up rappels, hand lines, backup harnesses, and for numerous other applications. Although webbing comes in various sizes and thicknesses, climbing-grade 1-inch tubular webbing works the best for multipurpose canyoneering.

A sling or runner is webbing that is tied or sewn in a loop.

You will need varying lengths of webbing for canyon descents. For rappels, webbing is tied in a loop around an anchor such as a tree, chockstone, or log. A small tree or bolt may require a 6-foot piece. A large chockstone or a tree some distance from the edge of the drop may need a 10- or 20-foot piece. It is best to carry several pieces of each length between your canyoneering group. In addition, each canyoneer should carry a 25- or 30-foot length or longer for hand lines, short rappels, a backup harness, or emergency use. This long piece can always be cut to shorter pieces for rappel stations if you run out of short pieces.

Accessory cord is most often used for pack lowering, to make a retrieval line, or to use as a hand line when crossing a stream or making a traverse across a ledge. Six-millimeter cord is perhaps the best all-purpose cord because of its high strength-to-weight ratio. One or all in your party should consider carrying 50 to 100 feet of accessory cord.

Climbing Protection

Rock climbing and advanced rappelling require that you use permanent or removable hardware to protect against a fall.

Permanent anchors generally involve a piton or expansion bolt. Pitons are metal spikes of various sizes that can be hammered into a crack. Expansion bolts are hammered or screwed into a drilled hole and then usually cemented into place. A complete kit of climbing protection consists of a set of bolts, a set of pitons, a hammer, a drill, an Allen wrench, cement or glue, hangers, washers, a blow tube, a wire brush, rings, and chains.

Removable devices come in many forms. Chocks are designed to wedge in a crack, crevice, or hole. Spring-loaded camming devices come in numerous sizes. To use one of these pieces of hardware, the climber retracts the device, places it in a crack, then expands the spring-loaded lever. It is removed by retracting the levers.

ELECTRONICS

The use of electronics is well established in the wilderness. Most canyoneers carry a camera, headlamp, and watch. Cell phones and global position systems are becoming more commonly used in today's canyons.

Cellular Phone

Cellular phones are becoming more common in the wilderness, although there is much debate over their place in the outdoors. Some people believe that having a cell phone gives less-experienced wilderness travelers a sense of security that leads them into situations beyond their skill level. Others maintain that cell phones facilitate rescue; a quicker response can minimize dangerous and costly rescues.

It is important to remember that cell phones are not always reliable. They can be rendered nonfunctional if they get cold or wet and when the battery dies. They often don't work in deep canyons.

Global Positioning System Unit

Global Positioning System, or GPS, units, are hand-held instruments that locate your position by using navigational satellites. Like cell phones, they can be rendered useless by cold, moisture, or dead batteries. Also, they have blackout areas, which usually include deep canyons. More importantly they require a lot of skill to use. You should always carry and know how to use a map and compass even if you have a GPS unit. Carry extra batteries also.

CAMERA

On most canyon trips, you will want to bring your camera along. Some ultralight canyoneers forgo taking along a camera due to weight considerations. If you will be taking photographs, keep in mind that the light is low in deep, dark canyons and you may have stretches of little or no direct sunlight. Also, the shadows and direct light may be difficult to adjust for. Try to avoid mixing direct light and shadow in one picture.

For low light, use daylight film. High-speed film may make shooting in low light easier, but you will sacrifice some sharpness and quality to your photos. A lens with a large aperture will allow more light also.

Use a tripod and low-speed film when taking landscapes in deep canyons. The tripod helps steady the camera, which is important when using a slow shutter speed. Some people get by with a monopod or a hiking staff that converts to a camera mount. Another option is to take a small, portable tripod, which you can find at some camping stores. A flash can lighten shots of people close up but doesn't help much with scenics or distance pictures.

GEAR MAINTENANCE AND REPAIR

Regular attention to maintenance, cleaning, and repair will prolong the life of your equipment and minimize problems while you are deep down in a canyon. Thoroughly inspect your equipment before you pack for a trip and after you return. Canyoneering is extremely hard on gear. Abrasions, cuts, ultraviolet light, water, and extreme hot or cold temperatures can increase wear and tear. Check your tent, sleeping bag, drybags, backpacks, and especially shoes for tears in the fabric and seams. Check buckles, laces, clips, poles, and zippers for smooth operation. Remove batteries from electronics when they are not in use, and get fresh cells for each trip.

On all hardware and technical equipment, check for stress cracks, heat bending, or dings. Retire them before they malfunction, especially if they have been dropped or significantly damaged. Inspect the entire length of a rope for cuts and abrasions. If you have any question about the durability of a piece of equipment, retire it.

Wash all your stuff when you get home—especially those pieces with moving parts—and allow everything to dry well. Wash soft goods according to the manufacturer's recommendations. Clean your stove and water filter after each trip. Climbing hardware with moving parts such as carabiners and ascenders should be rinsed with water. Wash your ropes in accordance with the manufacturer's specifications.

Eventually something will break in the field. Learn how to do basic repairs on your equipment. Carry a few feet of 2- or 3-mm cord to repair a shoe lace, pack strap, or pole leash. A rubber strap that stretches is especially useful for lashing gear or strapping parts together. Safety pins or wire can be used to repair zippers, buckles, or straps. Duct tape, of course, can be used for almost any repair. A multi-use tool or pocket knife can also be indispensable for making repairs. Some people carry a small patch kit to repair holes in a backpack, tent, or rain gear.

If you are biking you should carry a repair kit to fix a flat tire, spoke, break, chain, or derailer. If you are traveling by kayak or raft, you should carry basic repair kits to fix holes and leaks. You will probably be driving to the trailhead; a list of items to keep in your vehicle is included in Table 1 (pages 34–35). Your vehicle should be in excellent working condition and have regular tune-ups. Learn to change flat tires and to make basic repairs in the field.

PACKING

Packing can be part of the fun of your trip. Take some time and be methodical to make sure you don't forget anything. Assemble and check your equipment well in advance of your trip. Make sure gear is in good working condition. Some things, like film or batteries, can be picked up at the last minute. Others, like a new bulb for your headlamp or parts for your stove and water filter, may be more difficult to find and you may have to order small specialty parts. Try to replace broken or worn parts soon after you get home from a trip so your gear is ready for the next outing.

When you assemble your gear, sometimes it is helpful to lay it out on your garage or basement floor. This helps you to make sure you don't forget something, and makes packing easier.

Pack your equipment in drybags for wet trips or in stuff sacks for dry trips. Use one for spare clothes and overnight gear, another for a sleeping bag, and a third for food and essentials. With your supplies in separate bags, it will be easier to find things without unpacking you entire backpack.

Load your backpack so it is comfortable, balanced, and stable when it rides on your back. Gear should be organized and easy to reach. You don't want to unload your entire pack to get a water bottle, food, sunscreen, or rain gear. In general, heavy, bulky things should go near the back and at the bottom of the pack. This lends stability and keeps your center of gravity lower. Put your sleeping bag and tent at the bottom. Overnight food, stove, and camp clothing goes in next, near your back. Anything needed for the day should be closer to the top or in inside or top pockets.

Also, try not to load your backpack too high or close to your head. This may make it difficult for you to look up when you are wearing it, as the bulk will hit the back of your head.

Use the compression straps built into the side or front of the pack to cinch the load. This evens the load on the pack and keeps things from flopping around. If you must tie anything on the outside of your pack, make it your closed-cell foam sleeping pad. Be advised that canyoneering is so hard on gear that anything on the outside of the pack will likely get trashed. Make sure everything is secure, especially water bottles. Anything clipped or strapped to the outside can get torn off when you are canyoneering in tight slots or turbulent water.

One bit of advice is worth repeating: Don't forget to take extra gear, especially food, water, and clothes, even if some of those items must stay in the car for your trip. Sometimes the weather changes the morning you set out from your car; you may find at the last minute that you need rain or sun protection that you don't want to have left at home. When you get to the trailhead you can put the finishing touches on packing. Sometimes, especially on lightweight trips, you will be sharing gear. Divide up group gear such as stove, water filter, tent, ropes, first-aid kit, and emergency gear at the trailhead.

4

NAVIGATION

Navigation is part of the challenge of canyoneering. Orienteering, or map and compass navigation, is a rigorous competitive sport in itself. Sometimes navigation is fairly straightforward. On entry-level trips you may simply hike down or up a canyon from one point to another following a map and trail. However, the more advanced the route, the more you will need to rely on navigating using a map, compass, and altimeter. Map and compass navigation is an essential skill for all wilderness travelers, including canyoneers. In fact, canyoneering rescues are very often undertaken because travelers have gotten lost.

Every member of a group should learn how to read a topographic map and learn the two basic functions of a compass: to pinpoint his or her location on a map and to find the way to another location. If you become lost or you need to confirm your location along a canyon, pinpoint your position by taking compass readings on known land formations, then plot them on the map. Finding your way with a map is useful if you can't see where to go because of fog, brush, canyon walls, or other obstacles to following a trail or route. You can take compass readings from your known location on the map and then follow a route using your compass as a guide. Sometimes you will need to do both: locate your position, then follow your compass.

Map and compass use requires a great deal of practice. If you are off by only a few degrees you can totally miss the road or ridge you were aiming for. The basic principles for finding your way and identifying your position are outlined below. Keep in mind that this chapter is only an introduction to canyon navigation. Several books are available that detail navigation; see Appendix 2 for a list. Take a course on orienteering from your local college, mountaineering club, or guide service. Learn this skill and practice often. You can also train yourself near home; you don't need to be in a canyon to use your map and compass.

MAPS

General maps, as discussed in Chapter 2, come in all types and scales. The United States Forest Service, National Park Service, and Bureau of Land Management have maps covering large areas, but these are usually not detailed enough for finding your way or identifying your position.

A routefinding map is a topographic map, or topo. A topo has topographic or contour lines that mark particular elevations and are connected to other lines of the same elevation. Steep sections, such as canyon walls or cliffs, are marked with lines close together. Wide, flat streambeds are shown with topographic lines far apart. Lines that are formed in a circle, close together, and that show ascending elevation signify a prominent peak or plateau; circular lines that show descending elevation signify a depression such as a dry lake. Reading a topo takes some time and practice. The trick is to be able look at a two-dimensional map and imagine the three-dimensional land formations in the area depicted.

In addition to topographic lines, these maps have lines that correspond with direction. There are east-west or longitudinal lines and north-south or latitudinal lines or meridians. Longitude and latitude correspond to coordinates on earth. Coordinates are delineated by geographic measures called degrees, minutes, and seconds (not the same as minutes and seconds used for time). With a longitudinal and latitudinal reading you can identify a point on a map with two coordinates, thus locating your exact position. This allows you to communicate your position to others using universally recognized coordinates.

The standard topos in the United States are produced by the United States Geologic Survey (USGS). The USGS 7.5-minute series maps have a scale of 1:24,000. These have the most detail of any maps and cover the smallest areas. The USGS has some larger topos called quadrangles, but these have smaller scales and are thus not as useful for routefinding. Wilderness area maps or maps made by private companies may have topographic lines and, depending on the scale used, can be used for routefinding. Other countries have comparable topographic maps as well.

Well-researched route

Maps can be obtained from many climbing, mountaineering, or camping stores. To order topographic maps for the United States directly, contact USGS Information Services, PO Box 25286, Denver, CO 80225; 800-USA-MAPS. Topos can be ordered from the USGS website

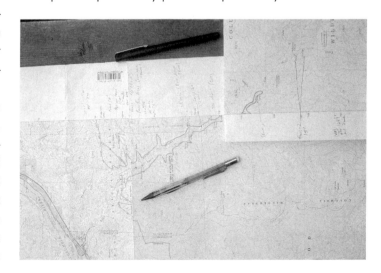

on the Internet at *www.usgs.gov* or *mapping.usgs.gov/esic/esic.html.*

Another option is to use CD-ROM maps, if they are available for the area you plan to visit. These maps are useful because you can draw in the route your party will be taking, or mark water sources, landmarks, and campsites in your area of travel. When you are finished customizing your route, you can print out a copy for everyone in your group, provided you have a high-quality color printer.

When looking at a topographic map, always check the scale. USGS 7.5-minute maps have a scale of 1:24,000 which means 1 inch is 2,000 feet, which is slightly less than half a mile. This will help you gauge hiking distance and time. Check other data listed on the map. These usually include a date the map was last updated and the local magnetic declination, which is discussed below.

You may need to take information from other maps and update your topo, writing current facts on the topo or correcting outdated data yourself. It is especially important to record current information about trails, trailheads, roads, parking areas, buildings, dams, and ranger stations. Mark major landmarks and obstacles like waterfalls or arches if they are not already on the topo. Make notes about seasonal landslides, log jams, or boulders. Put dates next to your notations so, as things change over the years, you have a record. During your hike, consult your map often. Stay on your planned route except when you must make alterations as needed for safety. Make notes on your map during the hike as well. Mark obstacles, exit points, seasonal changes, and other information you find out about en route. If you need to backtrack, if you make the trip again, or if you know others who will, this can be important information. Take frequent compass bearings and altimeter readings as discussed below so you can follow your progress. Keep your map handy but protected. Store it in a re-sealable plastic freezer bag to keep it dry.

COMPASS

The compass is a standard tool for navigation, but one that takes practice before you become completely efficient using it. The compass has three basic parts: index arrow, dial, and magnetic needle. The arrow is used to aim the compass at a landmark to take a bearing or follow a compass reading. The dial is a protractor with degrees marked to help you find your bearings. The needle always points to magnetic north.

Although the needle—and so all compass readings—is based on magnetic north, maps are not. True north, the direction of geographic north pole, is the basis for all maps. This difference between true and magnetic north is called declination. Since the direction of magnetic north changes depending on where you are around the world, so does declination.

Several methods have been devised to correct for declination. The easiest and perhaps most foolproof is to spend a few extra dollars and get a compass with a declination correction built in. These compasses are adjustable, and can be set to the declination for the area where you are canyoneering. The more traditional method used with a basic compass to adjust for declination is to manually add or subtract declination from your readings. Some find it is easier and less expensive to buy a basic compass and mark the declination on the compass with a piece of tape or using an ink pen. Whatever the case, learn how to adjust for declination, as it can

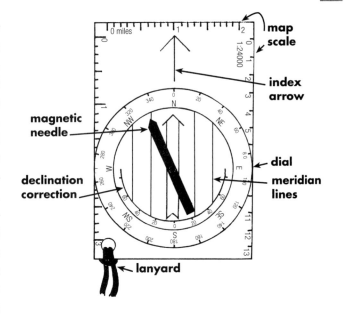

make a huge difference in knowing your bearings. A mistake in declination correction can be costly, especially if you are lost.

MAP AND COMPASS NAVIGATION

As mentioned, there are two main uses for a map and compass: to identify your position and to find your way.

Identifying Your Position

If you are lost or unsure of your location on your route, you can identify your position on the map.

1. First visually identify a prominent geographic landmark such as a pinnacle, plateau, or other feature. Then locate it on the map using contour lines, canyon rims, roads, or streams to help. Once you have positively identified the landmark both in the field and on the map, take a field bearing, which is a bearing based on magnetic north. Aim the arrow at the landmark and turn the dial so that the needle and the dial point to magnetic north. With the needle and dial pointing to magnetic north and the arrow aimed at the landmark, read the bearing at the arrow.

2. Next, correct your bearing by adding or subtracting declination. Turn the dial clockwise to subtract a west declination. Turn the dial counterclockwise to add an east declination. Again, if you have a declination correction on your compass, this step is unnecessary.

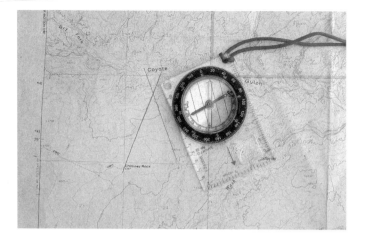

3. Then plot the bearing on the map. Put the edge of the compass on the same landmark as you have visually identified as it appears on the map. Rotate the compass without moving the dial so that north on the dial points to north on the map. Draw a line on the compass edge adjacent to the landmark. Your position is along this line.

In most situations, you will need to take two field bearings. Repeat the above steps with a second landmark that you have identified both in the field and on the map. When you plot the second bearing on the map, the point where the two bearings intersect on the map is your position. This process of taking two bearings and plotting location where they intersect is called triangulation.

In canyoneering, you can use the canyon floor or streambed as your second reference line instead of plotting a second bearing. After you plot your first bearing, the point where it intersects the canyon is your position. The process of using one bearing and the stream as a second reference is called bisection.

Upper: Triangulation using two field bearings

Lower: Bisection using one field bearing

Finding Your Way

To find your way in limited visibility or in complex terrain, choose your desired direction of travel on a map and then follow the compass reading using the steps outlined above, but in reverse order. To find your way, you must know your position on the topographic map, so keeping track of your progress throughout your hike is important. You may need to first identify your position using bisection or triangulation before finding your way, as discussed above. The following are three steps to follow to find your way with a map and compass.

1. Take a map bearing based on true north by aligning your compass on a map, with the edge of the compass touching your current location and the spot you want to hike to. Your compass arrow should point in your desired direction of travel. Second, while holding the compass, turn the dial until it points to north on the map. Read the map bearing shown on the dial next to the arrow.

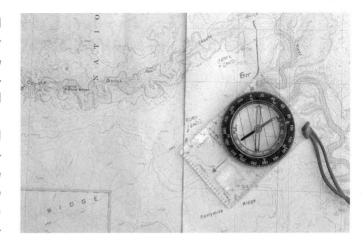

Taking a map bearing

2. Correct the reading by accounting for declination. Turn the dial counterclockwise to add a west declination. Turn the dial clockwise to subtract an east declination. If you have a declination correction on your compass, use that instead of converting your bearing.

3. Follow the reading by holding the compass in the palm of your hand and turn the entire compass so the needle and the dial point to magnetic north. The arrow shows your direction of travel.

You may not be able to travel directly to your destination due to obstacles such as water or canyon walls. You may have to take a bearing on an interim landmark, travel to it, and take another bearing. If this is the case, you may have to leap frog, making several short legs and finding your bearings several times before reaching your final destination.

Wrist watch altimeter

ALTIMETER

An altimeter measures elevation by recording barometric pressure and then calculating altitude. As you gain elevation, atmospheric pressure decreases; as you lose elevation, pressure increases. As you hike, the difference in pressure is detected by the altimeter and converted to elevation. An altimeter can assist you when you are locating

your position on a map. Mountaineering altimeters are perhaps the most accurate. Several outdoor wristwatches have altimeters built in; they are useful for general purposes, giving you a rough idea of your elevation gain over the course of a day. However, they are only accurate within 20 to 100 feet of your actual elevation, depending on the model.

Since an altimeter is a barometer, weather as well as elevation change affects your

SCOUTING

I've been scouting the route for two years. Drawn to the canyon to explore routes closer to home, pioneer wilderness travel the way the ancient ones did. But still the canyon eludes me; I'm not quite ready to make the drop.

The canyon has a popular rim hike, probably one of the most popular hikes in the Columbia River Gorge. It follows the rim a good 10 miles, crossing and re-crossing the stream. But to hike, or rather canyoneer, down the streambed below the rim is to enter another world, one with cold water, slippery rocks, tight narrows, and tall, sheer cliffs. Sometimes the rim trail gives access to the stream, but most times you look down into a deep, dark, inaccessible slot. In a few spots waterfalls make the canyon impassable without the use of technical canyoneering skills.

I first hiked the rim trail years ago as a kid, and visited it a few times thereafter. It wasn't until I began canyoneering in the American Southwest that I revisited this canyon in different terms: scouting a descent down the slot canyon. I had the topo maps and studied them over and over. I hiked the rim trail with my wife when she was four months pregnant and then again a year later with little Skylar bundled in the backpack baby carrier. On each trip we walked down to the waterfalls to inspect the rappels and take photographs, discussing potential anchors and escape routes.

I made several solo scouting trips directly up the streambed. The going was tough as I hiked upstream on slippery boulders and through swift water ranging from ankle-deep riffles to deep wade pools. I tried to hike on the bank, but with thick underbrush the going was too slow. Occasionally a fallen log gave me a quick path across boulders or rapids. At one point my leg got caught up in a fishing line. I slipped repeatedly and braced myself with my poles.

After several more hours of tough hiking I was stopped by a large plunge pool and a waterfall. Getting up the dry rock alongside the waterfall would be a 5.7 rock climb. But before getting on the rock I'd have to swim across the plunge pool

reading. You can compensate for the weather changes in pressure by calibrating the elevation in your altimeter at a known elevation. For example, set your altimeter at the parking lot, at the rim to the canyon, and at camp if you know your elevation based on a topo. As weather changes overnight, you will probably want to record the altimeter reading at night, then reset it in the morning.

For instructions on using an altimeter in weather forecasting, see Chapter 5.

and use rock climbing moves to climb out of the water. But I stopped at the pool. There were many reasons I cited for not continuing on: too late in the day to advance into a technical slot, no dry bag for my camera, and no wetsuit. The real reason I didn't continue on is because I was alone and afraid to swim across the pool and climb up the waterfall. Years in the wilderness have taught me respect, and this canyon will still be here next month, for another scouting trip, and next season, when the water is low and descent is safe. Alas, I will return again.

Scouting, Columbia Gorge, Washington

GLOBAL POSITIONING SYSTEM

Global Positioning System, or GPS, units use navigational satellites to give you a readout on your location and direction of travel. They can lock on to three satellites to give you a location reading on a built-in map, or they can lock on to four satellites and provide you with details of both location and elevation. They give bearing and distance to or from a programmed landmark. Be advised that using a GPS unit takes much practice, must be programmed, and is to be used in conjunction with a map, so you still need to be proficient at map reading. Fortunately, GPS units can now be connected to CD-ROM topo software, which was discussed earlier.

GPS units can be fast and reliable for navigation, but they do have some disadvantages. The units rely on batteries for power, and they can be damaged by cold or water. If you are deep in a canyon, your GPS unit may not be able to reach a satellite. See Appendix 2 for books that offer more detailed instruction on GPS use. Always carry and know how to use a compass for backup.

GETTING LOST

Eventually nearly everyone who spends time in the wilderness needs to find their way with a map and compass. Sometimes a quick look at the map or the simple use of a compass to find one's location is all that it takes. But sooner or later you may become lost. If you do, stop and regroup. Get some food and water and assess your situation. Then systematically use your map and compass skills, altimeter, GPS unit, and knowledge of the trail and route to find your way. Often, a short backtrack is all that is needed. At other times you may need to exit the canyon and travel along the rim. Don't attempt a route blindly. Take your time to determine where you are and where you need to go.

If it is close to dark, find a place to camp or bivouac. Most canyons don't lend themselves to night travel, and routefinding in canyon country is very difficult. You should have enough emergency food, water, clothing, and shelter for an unexpected night out.

Remember also that you can't count on a rescue. Search and rescue from outside parties is often delayed due to bad weather, difficult terrain, lack of personnel, and logistical problems. Self-reliance and self-rescue are the only things you can rely on, so be prepared.

5

WEATHER

Understanding weather is vital for canyoneering. It can make the difference between having a great trip or not going at all. However, to be good at evaluating and forecasting weather, you need practice, experience, and help from a variety of resources. The study of weather, or meteorology, is complex and difficult to master. This chapter is a general introduction to the subject and highlights weather hazards and good sources for obtaining weather forecasts.

A few important tips are worth noting at the outset. First, you can't just listen to one weather report and then head out. You must take into account the typical weather patterns of the area you are traveling to as well as the recent weather history. Learn the current seasonal variations as well as changes from recent years. Second, get a good handle on the current weather, both regional and local. Nothing is worse than making a long drive late into the night to find that a storm forestalls your trip. Third, get details on the forecast from reliable sources. Look at recent weather patterns, the current day's weather, and the forecast, and you should have an idea of the outlook for your trip. Finally, be ready for unexpected weather. Even the best weather predictors can be wrong. Always have an alternate plan, even if it means canceling your trip at the trailhead or making alterations in the route.

GENERAL METEOROLOGICAL CONCEPTS

Before trying to understand specific weather hazards and forecasting, it is a good idea to have a general comprehension of major weather phenomena.

Pressure and Fronts

The earth's atmosphere is that layer of gasses that covers its surface and "creates" what we think of as "weather." Pressure is the amount of force a mass of air in the atmosphere exerts on surrounding air masses. High-pressure systems are usually associated with stable air and good weather. Often high-pressure systems are composed of warm and dry air. Low-pressure systems are usually made up of unstable air and are associated with cool, wet weather. There are exceptions, however, and a warm low-pressure system, called a heat low, and marine air patterns can be cooler than high-pressure systems.

Pressure systems are noted on weather maps by isobars, which are lines that connect air of similar pressures. Isobars show how big a mass of pressure is and where it is moving. A high-pressure isobar is often called a ridge, and a low-pressure isobar is called a trough.

When one air mass collides with another, the meeting point is called a front. A cold front is a cold air mass pushing against warm air. A warm front is warmer air approaching cold air. Occluded fronts are combinations of cold, cool, and warm air masses. A stationary front is formed when a cold and a warm air mass are adjacent to each other but not moving.

Often in the field, one can feel the passing of a front or a change in pressure systems. Usually this is significant for sudden change in temperature, wind patterns, clouds, or humidity.

Humidity

Humidity is the amount of water vapor, the gaseous counterpart to liquid water, in the air. The percentage of water vapor in the air compared to the point of complete saturation is referred to as relative humidity. At complete saturation, or 100 percent humidity, the air condenses and forms clouds or fog. At 20 percent humidity the air is dry, and at 80 percent it is moist and muggy.

The dew point is the temperature at which air is at 100 percent relative humidity. Warm air holds more water vapor than cool air, so relative humidity changes with the temperature. For example, on cool nights, the temperature drops below the dew point, and fog develops. As the day warms up, the fog "burns off" as the temperature rises above the dew point.

By knowing the temperature and relative humidity, you can get a general idea of the current weather in a particular area. Here are some general pointers:

- High relative humidity and high temperature often combine to create moist muggy air.
- Low relative humidity and high temperature often combine to create warm, dry, and clear weather.
- Low relative humidity and low temperature can combine to create cool but dry air.
- High relative humidity and low temperature can combine to create cold, wet, rainy weather.

Wind Chill

Wind is movement of air from one place to another. It can move from areas of high to low temperature, or from areas of high to low pressure. There are some global wind patterns, such as the jet stream or trade winds, that are caused by the uneven heating of the atmosphere from the sun. Local winds are common in canyons, as discussed below.

Wind chill is the effect of wind and ambient temperature on exposed skin. For example, if the temperature is 40 degrees Fahrenheit with a 10-mile-per-hour wind, the effect on exposed skin is the same as an ambient temperature of 30 degrees.

Freezing Level

Elevation and land mass have a significant effect on the weather, especially the temperature. The temperature falls roughly 4 degrees Fahrenheit with every 1,000 feet gain in elevation, although clouds and other factors influence this rate. The freezing level is the elevation at which the air is 32 degrees Fahrenheit (0 degrees Celsius).

LOCAL HAZARDS

Uneven terrain, elevation changes, distance between pressure systems, the presence of bodies of water and vegetation, and other factors play a large role in weather. Local hazards are important to guard against, and you should always be prepared for a variety of types of weather even on a one-day trip.

Temperature

Cold temperatures can cause hypothermia or frostbite. Hot temperatures can cause sunburn, heat exhaustion, dehydration, and heat stroke. Be prepared for cool nights in the fall, winter, and spring.

When you are canyoneering, temperatures can change significantly during the day. Often the temperature on the rim will be hot, but deep in the slot it may be cold. Always bring clothes for sun protection as well as for cold weather. Chapter 6 further details what to do in cases of heat or cold injury.

Precipitation

Depending on the season, be prepared for precipitation, which may be rain, snow, sleet, hail, showers, or thunderstorms. In all seasons, carry a rain jacket and some type of shelter. Always have a dry change of clothes.

Stream Flow

Water flow can be variable and difficult to predict. You should know the drainage pattern for the canyon you are traveling through. If it has a large drainage system, storms and runoff from miles away can effect the water level. If you are in a watershed controlled by a dam or reservoir, check the water release patterns and know if any additional changes will occur when you plan to go canyoneering.

Keep in mind that water flow and precipitation can cause mud, mud slides, and slippery rock, so take extra caution.

Flash Floods

Thunderstorms can occur suddenly. When rain comes quickly and in large amounts, the ground sometimes does not have time to absorb it, especially in desert areas with hard, dry soil or slickrock. The result is a flash flood, a sudden flooding of canyons and streams. Keep in mind that rain miles away can cause a flash flood if the canyon you are in has a large drainage area. This can be the case despite clear, warm skies above the canyon.

Take extra care to watch for flash flood predictions or if poor weather is at all a possibility. In the desert southwest, flash floods can occur late in the day, even when you have enjoyed nice weather since morning. If you are down in a canyon for several days, you won't be able to see weather coming. Make sure the prediction is for clear skies overhead and over the entire canyon drainage for several days with minimal chance of thunderstorms in the afternoons. If you are hiking in a potential flash flood area, pick the time of year with the lowest potential.

When in the canyon, travel on high ground above the water line as much as possible. Sometimes you can see the high water mark, perhaps a demarcation between rock and vegetation or a log stuck in a slot. Always watch for signs of an imminent flash flood. The stream may rise, the water may become muddy, and you may hear a roar. Distant thunder may be an early sign.

The easy way across, Columbia Gorge, Oregon

If you are caught in a flash flood, stay as high and dry as possible. Don't forget to camp high and dry out of the flash flood zone.

Fog

Fog can make it difficult to see hazards, so use extra caution and hike slowly. Often early-morning fog or mist will burn off later in the day. You may need to wait a few hours until the fog lifts before you begin your hike.

Wind

Canyon winds can blow throughout the day, every day, all year long. During the day, wind blows up canyons and is strongest at the rim. In the evening, as the upper air cools, the winds blow down canyons or valleys. Canyon winds, especially the downdrafts in the evening, can be cool despite warm temperatures during the day. Take extra clothes for the evening.

Lightning

Lightning can occur almost anytime of year, often during winter storms or summer thunderstorms. If you get caught in a lightning storm, get to low ground and wait it out. Try to be lower than trees or peaks around you. Lightning can still reach you in deep canyons. Try to find a protected cove or a thick grove of trees or a deep slot. Stay out of the water and well above the high water line and areas where there is flash flood danger. Use your sleeping pad or pack to insulate yourself from the ground, and keep your feet close together. Avoid contact with metal objects such as canyoneering hardware.

FORECASTING

Forecasting the weather is one of the more difficult and uncertain tasks when you are canyoneering. Not only does it take some skill, but you may be relying on someone else's forecast. And weather prediction is not an exact science. Rather, it uses known facts coupled with the natural history of weather patterns to come up with a best estimate of what the weather will be. When putting together a forecast, always get details from a number of reliable sources.

As mentioned, you should first know the weather history in your area. You should know typical weather patterns as well as seasonal changes. These can be found from guidebooks, weather books, and weather charts. For example if you are in the southwestern United States, you should learn that the driest month is June and that the typical thunderstorm season is July through September.

Second, get as much information as you can about current weather conditions. Know if it is raining or cloudy locally or up canyon, as rain in these areas could indicate a chance of flash floods or high water in the canyon. Find out about an area's temperature, pressure, humidity, freezing level, dew point, and wind speed and direction. Get any updates on pressure systems or fronts nearby. Watch the weather year-round so you can put the seasonal variances together with typical patterns. Put all this together and make your best prediction at the weather for the duration of your trip.

General Sources of Information

Newspapers, television, radio, and the Internet can provide you with general forecast information. Television and the Internet can be particularly helpful, since they have satellite pictures and Doppler radar that shows local precipitation, fronts, and pressure systems.

The Internet offers much more than a quick glance at satellite or radar images, and you can get different views from different satellites or scroll back though several days, weeks, or months to see. The Internet can also give you immediate weather data from various sites. Some government or university sites have weather stations that give temperature, humidity, wind speed and direction, and other information that is updated frequently. Some have weather cameras that broadcast live views. There are so many websites that it can be confusing. Appendix 2 lists some of the general weather websites; visit one of these and navigate to your local area.

National Weather Service

The most complete weather source is probably the National Weather Service (NWS), a division of the National Oceanic and Atmospheric Association of the United States government. The NWS maintains a huge network of weather stations across the United States and has links to satellite or radar views. See the NWS website (listed in Appendix 2) for the most complete and updated information.

Another vital service from NWS is the weather radio. Using a small hand-held transistor radio that operates on several frequencies across the United States, you can receive the official forecast as well as current weather information twenty-four hours a day. The broadcast can be received in most urban areas as well as many rural areas.

You can also call the local NWS station by telephone to get the weather forecast and current conditions.

The NWS also puts out alerts for various hazards, including winter storms, flash floods, thunderstorms, tornadoes, freezing rain, blowing snow, and sleet. The service may put out an "advisory," "watch," or "warning." An advisory is a general warning that outdoor travelers should be extra cautious of. A watch means that the hazard is likely to occur within the next twenty-four hours or so. A warning suggests imminent danger.

Field Forecasting

After you have decided to go on a canyoneering trip, don't forget to continue to evaluate the weather. Look for changing cloud patterns on the drive out to your chosen location. Evaluate wind, temperature, humidity, and cloud changes constantly: at the trailhead, on the approach, in camp. You may need to change plans or cancel a trip due to changing weather. You should plan for an extra day in case of poor weather; you may have to wait out a passing storm or front.

Because barometric pressure changes with weather, an altimeter can also help with weather forecasts. If your altimeter reading rises without change in your elevation, this indicates a corresponding fall in atmospheric pressure. This can warn you of a low-pressure front that often means that a storm is coming. If your altimeter is steady or decreasing with no change in your elevation, this means the atmospheric pressure is steady or rising. Usually this signifies stable and possibly clear weather. Using the altimeter as a barometer is most useful when comparing changes over an entire day or overnight. Because both weather and elevation affect your altimeter, take readings and calibrate your altimeter in areas of known location and elevation, such as at the trailhead or camp. Check the altimeter reading before you go to bed and note any significant changes in the morning.

Forecasting Using Clouds

In addition to learning about fronts, pressure systems, and other factors, clouds are important, especially for field weather reading. A overview of clouds is provided in fig. 1. In general, whenever you have clear or scattered, thin, high clouds, the air is likely to be stable and good weather may continue. Clouds that are dark, thick, or accompanied by wind, can mean unstable air and poor weather. Other clues can alert you to changing weather. High, thin clouds at night or a halo of moisture around the moon can alert you to a storm coming in the next day or two.

Cloud reading can be complicated, but it is easy to learn the basic types of formations. Following are some general guidelines to identify different types of clouds.

- **Cirrus** are high, thin, wispy clouds that often signify warm, stable air higher in the atmosphere.
- **Cirrostratus** and **cirrocumulus** clouds have more moisture and are usually present higher in the atmosphere. They are also typically accompanied by stable, clear weather.
- **Cumulus** are white, puffy clouds that form with some moisture in the air. If they are sparse they are usually insignificant. But they can develop into rain clouds or thunderstorms.

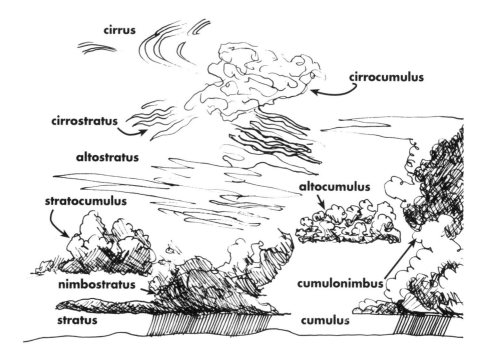

cirrus

cirrocumulus

cirrostratus

altostratus

altocumulus

stratocumulus

nimbostratus

cumulonimbus

stratus

cumulus

fig. 1

- **Altostratus** and **altocumulus** clouds are found lower in the atmosphere and are usually thick and gray. They can indicate worsening weather and significant moisture in the air.
- **Stratocumulus, stratus,** and **nimbostratus** are low-atmosphere clouds that are dark, thick, and usually bring rain.
- **Cumulonimbus** clouds, or **thunderheads,** develop quickly with moisture and unstable air. These produce thunder, lightning, and showers.

TO GO OR NOT TO GO

After you have studied the weather, you need to make a decision whether or not to go on your trip. You need to factor in location and duration of the canyon trip, your experience with and preparation for foul weather, and your ability to survive an unexpected night in poor conditions. On an entry-level canyon hike and with good preparation, you may be able to hike through a shower or two. However, on more advanced and technical trips, a cloudburst or thunderstorm could mean your progress is halted by a flash flood.

The more you learn about the weather, the more skill you will gain in predicting conditions for an outdoor trip. Consider reading a book listed in Appendix 2 or taking a community college course in meteorology to obtain more experience.

SNOWSHOE CANYONEERING

It had snowed and snowed all winter long. One of the thickest snow packs I could remember. But eventually, day-in day-out snowboarding got old. It was time to head to canyon country, to get away, to escape. I needed the canyon solace to rejuvenate my mind and body.

The winter canyon was quiet and uncrowded. There were a few people driving the rim road and nestling in the lodge's restaurant. The trail register showed that before the snow hit a late-fall party had made the two-day hike our group was about to undertake, but no one had traveled here in two months. The roadside was stacked with snow drifts. When we hitched out to the trailhead after leaving our truck at the trail's end, the scene was like one seen on a postcard: a thick, white blanket of snow covered the canyon, and the contrasting redrock pinnacles and the evergreen pinyon pines pointed to a deep blue, cloudless sky.

Once on the trail, we found that the snow was deep but easily manageable in snowshoes. The hike was quiet, with only the barest traces of wildlife—some tracks and scat—on the trail. On one traverse, a loud "woomfph" of settling snow startled us out of our zombie-like gait; we were walking along deep in our own thoughts hearing nothing but the steady, monotonous crunch of snowshoes. My stride was a bit shorter than those of my companions, so when climbing uphill it felt as if I was breaking trail, as I couldn't comfortably make the stretch into their tracks. On the downhills, we'd glissade on the tails of our snowshoes.

At night, dry air and body heat quickly dried our polyester clothing. We cooked instant soup, then squeezed into the tent and read by headlamp. We were asleep soon enough, though, exhausted from a long day.

That night I woke once and slipped outside to pee. The night air was near freezing and the sky was brightened by a million stars. The next day we would hike the canyon, climb up to the rim on the final push back to the car, then drive home. I'd be back to work the next day, and would probably be snowboarding the following. I dove back into my sleeping bag, too cold to stare at the stars for long. I was back to a deep, sound sleep in an instant. I didn't dream that night. Perhaps I didn't need to dream.

Descending into winter wonderland, Bryce Canyon, Utah

6

FIRST AID

Canyoneering is a truly fun sport: hiking through water, rappelling waterfalls, and squeezing though narrows sometimes feel like child's play. However, the wilderness is never risk free. Hazards exist around every bend, whether you are on an entry-level canyon hike or a technical descent. As mentioned in Chapter 1, taking personal responsibility for your safety is of paramount importance. Dangers can be minimized but never totally avoided. The level of risk you expose yourself to depends on weather and canyon conditions, the skill level of the people in your group, the difficulty of the canyon hike, the availability of outside help in an emergency, your preparation, equipment, and many other factors. Everyone in your group should be personally comfortable with an acceptable level of risk.

Field treatment of injury and illness requires formal instruction. Take a wilderness first-aid course such as Mountaineering Orientated First Aid (MOFA) or Wilderness First Responder (WFR) to learn how to treat basic injuries and illnesses common in wilderness travel. Also learn basic life support, or cardiopulmonary resuscitation (CPR). Check with your local American Red Cross or community college for courses.

This chapter provides a summary of the most common canyon injuries and illnesses. Remember that specific geographic or geologic features in your local area may warrant extra precautions. To prevent a medical emergency from arising, the following basic guidelines, discussed earlier in this book, should be routinely followed.

- Use the right equipment in good condition. Buy the best gear you can afford, and don't skimp when it comes to safety items.
- Plan and prepare well. Know the route and bring adequate food, water, and clothing. Choose a route consistent with your skill level.
- Stay in shape. This means undergoing both strength and endurance conditioning.
- Use good judgment. Human error contributes to many wilderness accidents. Be smart.
- Seek more information. Appendix 2 details books and selected Internet sites that

provide information about canyoneering. Mountaineering and rock climbing courses can give you some basic skills for technical canyoneering. Consider hiring a professional guide. Many are readily available for all levels of excursions.

FIRST-AID KIT

A first-aid kit is part of your essential equipment. Buy one from a reputable wilderness medical kit company, or assemble one of your own. The size will depend on the length of your trips, the number of people you travel with, your level of wilderness first-aid skill, and how far from help you expect to be. For day trips, a basic, compact kit with supplies for wounds is adequate. Table 2 lists some items for a basic first-aid kit. For long trips in remote canyons, you will need many more supplies.

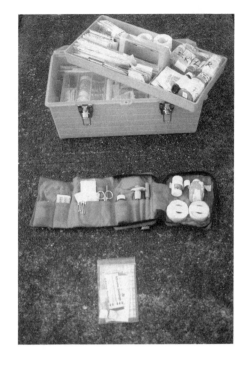

First-aid kits (top to bottom): expedition, overnight trip, day hike

TABLE 2

BASIC CANYONEERING FIRST-AID KIT*	
Small irrigation syringe	Waterproof first-aid tape
Antiseptic cleanser	Splint
Antibiotic ointment	Anti-inflammatory pain
Nonadherent dressing	medicine such as
Gauze pads, various sizes	ibuprofen or aspirin
Adhesive bandages, various sizes	Powdered oral rehydration salts
Butterfly bandages	Gloves and CPR microshield
Blister bandage such as moleskin	Safety pins
	Water purification tablets
Two-inch elastic wrap	Sunscreen
Cloth first-aid tape	Compact first-aid book

* This table lists items to be included in a first-aid kit, which is an essential item listed in Table 1 (pages 34–35), General Equipment for Canyoneering.

EVALUATION OF INJURY

When an accident occurs, always approach the situation in a step-by-step manner. This will keep your group organized and make first aid more efficient. The person with the most first-aid experience should organize the rest of the group.

1. Always prevent further injury: Only approach an injured person when you know you can do so safely.

2. Evaluate the person quickly and methodically. Sometimes you will need to do this at the site of the accident. Other times you may need to rescue the person from a precarious position and get to safe ground. In either case, always avoid doing anything that would cause additional injuries, either to the injured or to those rendering first aid.

3. In most cases, you can perform a primary survey by checking the airway, breathing, and circulation in that order (A, B, C). Make sure the throat is open and there is no debris in the mouth. Look at the chest and listen to see if the patient is breathing. Then check for a pulse in the neck and see if the skin is warm. If the patient is not breathing or does not have a pulse, CPR will need to be started promptly by a trained member in the party. If there is any chance of head or spinal injury, always immobilize the patient to prevent further injury. Again, this must be done by someone with proper training.

4. After the initial check, perform a secondary survey, or more detailed head-to-toe assessment of the body. Focus on the injured area, but check the entire body quickly for hidden injury. Check for bruising, deformity, bleeding, and pain that suggests a more severe injury. Always cover the injured person, as he or she is likely to get cold quickly, especially if wet.

5. Initiate first-aid treatment following the guidelines learned in a first-aid class. If you have a book to remind you of treatment, use it as a guide.

6. Make a plan for the rest of your trip. This should be a group decision based on the details of the situation. You may choose to simply treat a small wound and continue on with the trip. For more complex and serious injury or illness, you might have to head back to your vehicle. If an injured person can walk, you can make a safe and fast exit. Also consider that you may need to turn back, hike to the rim, or continue down canyon to make the quickest and safest exit. Oftentimes heading down canyon is much easier than trying to retreat up a steep, narrow slot. You should find out if there is a trail to the rim from which you can get back to the main trail or road much more quickly.

   If the injured is unable to ambulate, your group will need to make a decision whether to carry the person out or bivouac and get help. Again, make this decision as a group based on the seriousness of the injury, first-aid skills of the group, distance from help, expected weather, conditions of the canyon, time of day, equipment available for either evacuation or bivouac, and other factors. If you decide to bivouac, some people should go for help immediately and others should set up camp. If there are only two of you, you may need to leave your injured partner with extra food, water, clothing, and emergency gear and head out for help alone.

7. All injuries and illnesses should be evaluated by your doctor at the end of your trip.

WOUND CARE

Minor wounds are common and occur on many canyon hikes. All wounds, whether small or large, should be properly cleaned and bandaged to speed healing and prevent infection. To treat a wound, perform the following steps:

1. Find a place for the injured person to rest that is dry and protected.
2. Wash your hands and put on first-aid gloves.
3. Clean the wound with plenty of purified water—the more the better, so long as you have plenty of water available. (See Chapter 2 for methods to purify water; note also that many first-aid kits contain iodine tablets that can be used for this purpose.) Irrigation will remove germs and dirt and is one of the most important steps to take in preventing wound infection. The larger the wound, the more water you should use to clean it. Use a small syringe to help irrigate the wound. This will save water by directing it at the wound and add pressure to dislodge deeply imbedded particles.
4. Cleanse the area with antibacterial soap.
5. Spread an antibiotic ointment over the wound to keep it clean, prevent infection, and provide protection.
6. Cover wounds with a bandage. For small cuts, butterfly bandages will close wounds and speed healing. However, butterfly bandages can trap infection, especially in dirty wounds. If the wound is large, use gauze and first-aid tape as a bandage. Sometimes an elastic wrap over the primary dressing is needed to provide additional protection.

    Keep the dressing as clean and dry as possible. If you will be hiking in water, use waterproof first-aid tape. This plastic-coated cloth tape resists water, so it continues to stick when wet and usually keeps the wound drier than standard first-aid tape. Some people use duct tape for this purpose, but it can irritate the skin and it isn't waterproof.
7. Check the wound and injured extremity often. Look for signs of infection such as redness, heat, pain, pus, and swelling. The bandage should not hamper circulation. Always look at fingers and toes to make sure they stay warm.

BLISTERS

Blisters are common on any hiking trip. They can be prevented much of the time. Try walking barefoot around the house before the hiking season to toughen your feet. Keep your toenails cut short. Wear good socks and make sure your shoes or boots fit properly. Use custom foot beds if necessary and always break in new boots.

If you note a sore spot on your heel, take off your shoes and put a pad on the area. A small red spot usually forms before a blister does. If you get a blister, avoid popping it; infection is more common and you lose the natural protection of skin and fluid of the intact skin. If the blister pops on its own, follow the wound care guidelines given above.

For padding against blisters use moleskin or adhesive-backed cotton flannel. For larger blisters you may need to cut a circular hole in the first layer of moleskin, then use a second piece over it. Some people prefer nonadherent blister bandages, which come in several brands, or a liquid blister treatment that dries to a flexible coating. Whatever the case, you may need to put tape on top of the blister bandage for additional protection. Always pad your boot to prevent further pressure on the blister.

SPRAINS, STRAINS, FRACTURES, AND DISLOCATIONS

Sprains and strains occur in canyoneering especially as you are wading over slippery rocks, climbing up and over logs jams, sliding through chutes, and squeezing through narrows. A sprain is an injury to a joint comprised of ligaments and tendons. Sprained knees and ankles are perhaps most common with hiking. Knees can be banged up rappelling or down climbing chutes. Wrists are often sprained when you catch yourself with outstretched hands. A strain often refers to the stretching of or a minor tear to a muscle. Muscle strain with hiking is often to the legs, usually to the hamstring, quadracept, calf, or groin. Strain can occur when the body is

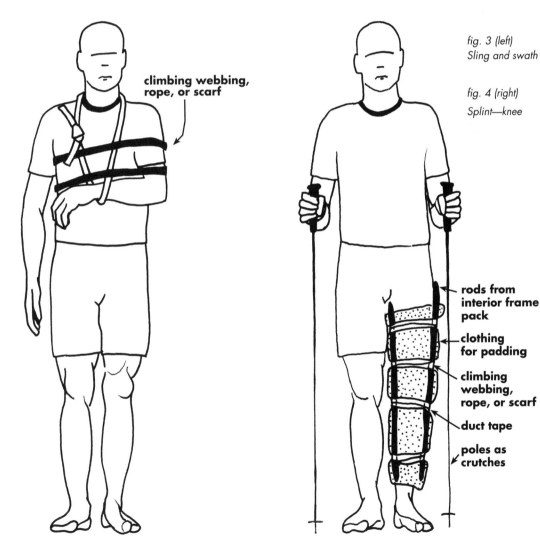

use Ace bandage or climbing webbing

small branch for splint

fig. 2
Wrist/forearm
splint

fig. 3 (left)
Sling and swath

fig. 4 (right)
Splint—knee

climbing webbing, rope, or scarf

rods from interior frame pack

clothing for padding

climbing webbing, rope, or scarf

duct tape

poles as crutches

not warmed up in the morning or not stretched after a long day. Sudden use of a muscle, for example when you catch yourself from a fall, can cause strain as well.

Fortunately, using hiking poles lessens stress on legs and helps to prevent falls. They are especially handy for hiking in streambeds. Good judgment minimizes risks as well: avoid jumping in plunge pools, sliding too fast down a chute, or bounding from boulder to boulder. Warm up in the morning and warm down at night with light walking and stretching.

It can be difficult to distinguish a sprain or strain from a fracture or dislocation. Discolored skin, swelling, a deformed bone, and tenderness can all signify a more serious injury. Bone abnormality, asymmetry between right and left limbs, the inability to walk on a sprained ankle or knee, and the inability to move an injured arm or leg all suggest a more severe injury. When in doubt, assume the injury is serious.

The initial treatment for sprains, strains, fractures, and dislocation is primarily RICE: rest, ice, compression, and elevation. It is best to rest for a few minutes, elevate the injured arm or leg, apply a cold water compress if available, and wrap the injured extremity. The goal is to minimize swelling, speed healing, and retain some function. Anti-inflammatory medications such as ibuprofen or aspirin will help with swelling and pain.

Reevaluate the injury after a few minutes. If the injured person can move and use the arm or leg without pain, it may be a minor injury. But know that even minor sprains and strains are susceptible to reinjury and will not be 100 percent recovered in terms of strength or mobility for several weeks. The injured person may need to lighten his or her load and let others carry some of their gear. For all but minor injuries, immobilize the joint with a splint and abort the canyon trip.

When splinting, use a lot of padding; T-shirts or foam sleeping pads work well. Use a metal splint from your first-aid kit or improvise with trekking poles or the removable metal stays from an internal frame backpack. Splint the entire forearm and hand in wrist injuries, using an Ace bandage or climbing webbing, as shown in fig. 2. Use a sling and swath to immobilize the arm for shoulder or collarbone injuries, as shown in fig. 3. For knees, splint the entire leg, as shown in fig. 4. For ankles, leave your boot on to act as a splint.

To secure the splint and padding, use duct tape, first-aid tape, accessory cord, webbing, or extra shoelaces. Make sure the splint is tight enough to keep the arm or leg still, but allows for circulation. Check the toes and fingers routinely to ensure they are warm.

DEHYDRATION

Dehydration occurs year-round, although it is most common with the warmer temperatures of spring and summer and in desert climates. Dehydration usually begins with headache, dizziness, fatigue, nausea, and cramps. Thirst is often a late sign.

To prevent dehydration, drink fluids regularly. Try to regulate your water intake according to need. Don't drink too little or too much. Try to keep your water bottle handy, not buried in your pack. Some people use packs with water bottle holders, but many bottles you can clip to your pack where you can easily reach them. Keep in mind that with the rigors of canyoneering, you can easily lose a water bottle that is not well secured to your pack. Consider using a hydration bladder if that style of water carrier works with your pack.

Watch your urine output. If you pee infrequently and your urine is scant and dark yellow, you may not be getting enough fluids. If you pee every ten minutes and your urine is clear, you are getting plenty of water and may want to scale back a bit.

Dress appropriately. In hot weather, keep covered from direct sunlight. In cold weather, try to stay warm but avoid overheating and generating too much sweat.

To treat dehydration, drink fluids. A sports drink can help your body rehydrate faster by replacing lost salts and sugar. In severe cases, commercial oral rehydration solutions are the quickest way to rehydrate. These are pre-measured packets of powdered electrolytes that are mixed with water. The packets are available in many first-aid kits. Avoid alcohol and caffeine; they are diuretics and cause dehydration.

COLD INJURY

Cold injury is a risk when you hike canyons in the fall, winter, early spring, or any time of year if you will be passing through a lot of water.

Frostbite, Frost Nip, and Trench Foot

Frostnip occurs when ice crystals form on the skin. It progresses to frostbite when the skin gets damaged from freezing. Early signs include numbness, pain, redness, and swelling. Sometimes a white patch of ice crystals can be noticed on exposed skin. Often frostbite on the face is noticed by one's companion. In advanced cases, the skin becomes firm, painless, dull, and pale. Trench foot is cold damage to feet that is similar to frostbite. It occurs after prolonged exposure to cold and wet, for example after hiking in frigid water for days.

With any cold injury, warm up and dry out. For frostbite and trench foot, the best course it to immobilize the arm or foot with a splint. Anti-inflammatory pain medicines like ibuprofen and aspirin will help relieve pain and decrease swelling. Avoid rubbing the frozen skin. Definitive treatment done by a properly trained person is rapid warming with nonscalding hot water. Perhaps the most important thing to remember is not to let a thawed hand or foot refreeze; this can be more damaging than leaving it frozen.

Hypothermia

Hypothermia is a deeper cold injury than frostbite that occurs when the body's core temperature falls. This happens anytime you are cold or are wet for a prolonged period. Lack of nutrition or hydration also contributes to hypothermia. Early signs include shivering and frigid skin. More advanced signs include stiff muscles, slow breathing, fatigue, and lack of coordination. Eventually shivering stops and a feeling of warmth replaces the sensation of cold. Major alterations in mental status, such as confusion or difficulty moving, are signs of advanced hypothermia. The person's sensations of feeling hot or cold and skin temperature are poor indicators in advancing hypothermia.

The treatment for cold injury is the same as prevention: warm up and dry out. Change into dry, warm clothes. Put on a hat, mittens, and thick, dry socks. Cover all exposed skin, especially the face; use a balaclava or neck gaiter. Move to generate heat: wiggle fingers and toes, swing arms, walk. If you stop for a rest, sit on an insulated pad or your pack. Eat high-energy, high-calorie food. Drink warm fluids and stay well hydrated. If the weather is severe, get inside a tent or sleeping bag. Build a fire as a last resort.

HEAT ILLNESS

Heat illness like cold injury can occur at any time of year, especially in the desert. Keep in mind that sometimes the canyon rim is much hotter than deep inside the canyon; you may have a long hike back to the car in direct sun.

Sunburn

Sunburn is common and, fortunately, easy to prevent. Since the condition is a delayed skin reaction, the burn is usually noticed hours after being exposed to direct sun. Usually sunburn is red and painful, but severe cases can lead to blisters and cracked skin.

Prevention is simple: cover up and use sunscreen. Wear lightweight long-legged, long-sleeved, light-colored clothes, and a wide brim hat to keep the sun off your skin. Don't forget to wear sunglasses as well. Covering up with clothing provides superior protection to using sunscreen, since sunscreen wears off with time, sweat, and water. Plus you will stay cooler with clothing blocking the direct radiation of the sun. If you have any exposed skin, use a waterproof sunscreen of at least 20 SPF (sun protection factor). Apply it liberally and frequently.

If you get sunburned, keep your skin clean and dry. Aloe-based moisturizing lotions may help healing and pain. Treat blisters promptly as outlined above.

Heat Exhaustion and Heat Stroke

Common to the desert and summer climates, heat exhaustion is a stress on the body caused by rising body temperature and dehydration. Rising body temperatures result from increased ambient temperature or internal heat generation from exercise. The dehydration is usually caused by sweating and lack of fluid intake. Initial symptoms include leg cramps, fatigue, dizziness, nausea, and thirst. The skin usually feels hot and dry, and the person appears dehydrated.

Heat exhaustion can quickly progress to shock from heat, also called heat stroke. Symptoms of shock include cool and clammy skin, confusion, lack of coordination, and nausea. Heat stroke is a true medical emergency; if left untreated, death can result. It is differentiated from milder forms of heat illnesses by a core body temperature greater than 102 degrees Fahrenheit and mental status changes.

To prevent heat exhaustion and stroke, stay cool. Avoid hiking during the heat of the day if possible. Seek shade and rest frequently. Time your hikes so you will be in open, sun-exposed terrain during early morning or evening. Wear a wide-brimmed hat and light-colored, loose-fitting long-sleeved shirts and long pants. When in the hot sun, consider periodically soaking your shirt in water to aid cooling. Acclimatization to the heat has also been shown to help prevent some problems. Plan on taking a few days to adapt to the heat if you are heading to the desert.

Immediate field treatment is vital to prevent mild heat illnesses from becoming life threatening. Fluids are the mainstay of treatment. Administering water with oral rehydration salts is the best course. Cooling should be initiated with fanning, misting, and wet towels. Once cooling has begun it is important to closely monitor the victim's body temperature. Since it is easy to induce hypothermia during the cooling process, cooling should be slowed or stopped when the patient's temperature falls below 102 degrees Fahrenheit. If you don't have a thermometer, slow cooling when the person becomes more alert, feels cool or of normal temperature to touch, and starts drinking on his or her own.

Water Intoxication

Water intoxication, or salt depletion, which is related to heat illness, also usually occurs in the desert or hot climates. It is differentiated from other heat illnesses by the fact that the victim is not dehydrated. Unlike heat exhaustion or heat stroke, water intoxication usually occurs after a person has had copious amounts of water but little or no intake of salt. Symptoms include nausea, headache, blurred vision, and fatigue. Generally, if someone you suspect of water intoxication has drunk and urinated within an hour of the onset of symptoms and does not look or feel dehydrated, he or she may have the condition.

The field treatment for water intoxication is similar to that for other heat illnesses: administer electrolyte fluids, such as an oral rehydration solution. Salt tablets should be avoided; they are too concentrated and do not allow for proper absorption into the blood stream. Salty foods do help replace lost salts.

NEAR-DROWNING

Drowning can occur in canyons with high water, large plunge pools, and flash floods. Near-drowning refers to the recovery of an unconscious person after submersion in water. Drowning or near-drowning are often accompanied by panic, struggle, breath-holding, and inhaling water.

If possible, stay high and dry. But since part of the fun of canyoneering is getting wet, be cautious. Scout the canyon well. Check weather and canyon conditions. Rappel to the sides of waterfalls and avoid the direct forces of rushing water when entering a plunge pool.

The primary treatment for a drowning patient is CPR. First remove the victim from the water and administer mouth-to-mouth breathing. Performing the Heimlich maneuver or doing abdominal thrusts on the patient in an attempt to remove fluid from the lungs is not beneficial.

ANIMALS

The wilderness is teeming with creatures that can bite, sting, and envenomate, but with some knowledge of prevention, most canyoneers can avoid animal attacks. With some exceptions, most animals are not aggressive toward humans unless their nests or territories are invaded or their young threatened. To avoid animal attacks, follow basic precautions.

Look before stepping or reaching. Shake out clothing, footwear, bedding, and equipment. Make noise, especially when you are in bear country, so you don't surprise animals on the trail. Observe animals from a distance and if they appear agitated, vacate the area. Avoid animal nests and feces. Practice caution when exploring dark canyon crevices. Store food, garbage, and waste properly. Observe good hygiene. Treat your water appropriately, as discussed in Chapter 2.

Some animals are venomous or harbor microorganisms that cause illness. Two categories of snakes, a few spiders, and one scorpion in the United States are potentially venomous. Young children and the elderly are particularly at risk. Some antivenins and immunizations are available for particular types of snakes, spiders, scorpions, and rabies; however these are usually administered in hospitals.

Consider using insect repellents such as DEET (N,N-diethylmetatoluamide) or permethrin. However, these are toxic and can have detrimental effects on humans. Natural repellents such as citronella can work as well. Perhaps the best way to avoid insects is to cover up. Wear high-top boots, long pants tucked into socks, and long-sleeved shirts. Closely inspect your skin regularly. Consider wearing a mosquito head net and hat. Consider adding a snake bite kit or tick extractor to your first-aid kit.

Snakes

Venomous snakes strike particular fear among wilderness travelers. In North America, coral snakes and pit vipers such as rattlesnakes are common in the wild. Field treatment generally begins with basic wound care to control bleeding, clean the wound, and minimize inoculation. Minimize circulation of the toxin by immobilizing the limb and perhaps the entire body. Some experts recommend using a light compression wrap around the bite. This is controversial: without proper training it may be too tight and become a tourniquet, cutting off blood flow.

Use of mechanical suction devices for snake bites is controversial since they likely remove little or no toxin. The plunger-type device is most effective when used within three to five minutes after the bite and left on for at least thirty minutes.

Spiders

Although nearly all spiders are feared in popular culture, few are actually venomous. Two common venomous spiders are the brown recluse and the female black widow. The latter has a red hourglass-shaped mark on a black body. Tarantulas are a broad category of spider and include funnel web spiders and more familiar varieties covered with fuzzy hair. Some tarantulas bite, while others rub their hind legs to flick barbed venomous hairs that penetrate the skin of people or prey.

Spider bites usually impart a sharp bite, which is followed by symptoms ranging from mild skin irritation to more systemic symptoms like fever, chills, nausea, vomiting, weakness, and muscle aches. A brown recluse bite can lead to local bruising and blackening of the wound. A black widow bite can cause neurological problems such as muscle spasms, rigidity, or weakness.

As for snake bites, the goal when treating spider bites is to minimize circulation of the toxin. The victim's limb should also be immobilized.

Good wound care includes inspection, irrigation, and cleaning with antiseptic cleanser to reduce inoculation.

Ticks

Ticks, which live all over rural and wild areas, usually attach to humans and pets in the summer months. They cause disease by either envenomation or by transmitting bacteria or viruses. In North America, several ticks are prevalent. The pajaroello tick puts out a toxin that causes local redness, pain, and itching. The bite can progress to paralysis of the extremity, which usually resolves with removal of the tick. The deer tick, the Western black-legged tick, and others can transmit the bacterium that causes Lyme disease. Lyme disease usually begins with a circular rash that enlarges with time. Rocky

Mountain spotted fever, whose symptoms include fever and a pink rash that progresses to red spots, is caused by another bacterium, which is often transmitted by the wood tick and the American dog tick. Colorado tick fever is caused by a virus transmitted by the wood tick. This fever usually comes and goes every two or three days.

First aid for tick-related illness begins with removal of the tick. Ticks often stay on the skin for several hours before burrowing, and after becoming attached they may wait several more hours before releasing venom, bacterium, or virus. Inspect your skin often and wear protective clothing, as noted above.

For removal, grasp the tick as close to the skin surface as possible with either first-aid tweezers or tick-removing forceps; the latter are available at many camping and mountaineering shops. Pull the tick with slow, gentle pressure to avoid crushing or dividing it. After removal, clean the wound thoroughly. Other methods to remove ticks, like using alcohol, fingernail polish, or heat, may induce the tick to salivate and are probably less effective.

Scorpions

Some scorpions, such as the bark scorpion in the southwestern United States, are venomous. The stinger on the end of the tail injects a toxin that produces local numbness and pain. Symptoms included nerve and muscle dysfunction, blurred vision, slurred speech, muscle spasms, pain, nausea, and vomiting. Field treatment is similar to that for spider and snake bites: minimize circulation of the toxin with wound care and immobilization.

Rabies

Rabies is one of the most feared animal-borne diseases. It is a virus that is usually contracted when humans are bitten by an infected animal. Bats, dogs, cats, raccoons, foxes, skunks, cattle, and other animals can carry the virus. Symptoms can begin days or years after a bite and initially include fatigue, anxiety, agitation, fever, depression, headache, nausea, vomiting, and pain. Subsequent neurological symptoms can follow two pathways. What is called the furious pattern involves agitation, seizures, and muscle spasms. The paralytic pattern presents symptoms like paralysis, lethargy, and lack of coordination.

If bitten, again follow good wound care including copious irrigation and use of antiseptic solution to reduce the virus inoculation. Try to observe the behavior of the animal and relay this to medical providers. An animal that bites is ideally, but rarely, detained by professionals for signs of rabies. In the wilderness, you need to assume that an animal that has bitten a person is rabid. The treatment is a rabies vaccination, which is available at a hospital.

Airborne Diseases

A number of airborne microorganisms exist in canyons. Hantavirus is a virus, transmitted by the deer mouse and other small mammals, that occurs in the southwestern United States. Animals pass the virus in saliva, urine, and feces, which are subsequently inhaled or ingested by humans. Symptoms begin with fever, cough, shortness of breath, chills, headache, nausea, and vomiting. They can rapidly progress to respiratory distress. Histoplasmosis is a microorganism found in caves contaminated with bat guano and bird droppings. Symptoms after ingestion include headache, fever, difficulty breathing, cough, nausea, vomiting, and chest pain. The microorganism coccidioidomycosis is found in dust, also in the Southwest.

Always use caution when exploring caves or dark, dry, dusty caverns. Minimize time and exposure to dust, especially in deep caverns or in windstorms.

Waterborne Diseases

Many bacteria, viruses, and protozoa exist in water. *Giardia lamblia,* one such common microorganism, causes diarrhea. Others include the fecal coliform bacteria, which are transmitted by cows, humans, dogs, and other animals feces. Rigorous good hygiene and treatment of all drinking water is the best prevention. Wash hands before meals and after bathroom visits.

Treat all water as described in Chapter 2. If you do get diarrhea, stay well hydrated with an oral electrolyte solution or water.

Other Critters

Many other types of venomous and nonvenomous animals exist around the world. Various bees, wasps, ants, flies, sucking bugs, beetles, and caterpillars can bite or sting, causing envenomation. Gila monsters, which are large lizards common to the American Southwest, can bite and envenomate, causing pain, swelling, weakness, dizziness, nausea, and chills. Follow traveling techniques that avoid contact with animals. Good wound care is always the mainstay of treatment.

PLANTS

Problems resulting from ingesting of poisonous plants occur less frequently than other wilderness illnesses. They usually take place only when a person misidentifies a poisonous plant for an edible one. Learn to be skilled at identifying edible plants in your area, especially if you are going on a prolonged trip where an injury or equipment failure could result in a prolonged bivouac or emergency situation. Consider taking a survival course or studying a survival book that lists edible plants if you plan extended trips in remote areas.

*Desert in bloom,
Esclante, Utah*

MAJOR TRAUMA

Major trauma can be disastrous and life threatening to the injured person as well as to the entire group. Avoid accidents at all costs. Follow the guidelines for safety and leave yourself a wide margin for error.

Major injuries fall into three main cat-

THE ROUGHEST ROAD

The road to San Borja was by far the roughest I'd ever traveled.

After a month on Baja California's west coast windsurfing, my wife and I decided to head inland and explore the desert arroyos. Driving south, we passed the turnoff for Bahia de los Angeles. A while later we came to the small village we were looking for and, with some searching, spied a rusted, hole-riddled, cockeyed metal sign: San Borja, 40K. By low-range four-wheel drive that meant three hours.

We drove up and down desert arroyos, among tall saguaro cacti. We drove over rock and dirt, thorns and sticks. Down a wash, we hit soft sand and had to rev the truck to keep from bogging down. Occasionally we saw signs of civilization: a fenceline or hacienda. There were a few cattle eating saguaro cacti that a local rancher had cut into pieces with a machete.

An hour into our drive, we saw another sign, as if strategically placed to keep us driving just as we got to a point where we were ready to wonder where we were headed. In another thirty minutes we passed a truck with two locals, smiling and waving, with a load of cacti for cattle in their pickup.

We bounced in our seats. Jennifer drove and I got out every twenty minutes to scout the road and guide her over rocks as we dropped down into the next wash. Later that day I found our roof rack had broken and was sagging—from all the gear—into the canopy.

We finally made it to the road's end, where there was a small paradise with a spring, a mission, and one house. San Borja Mission is one of the best preserved in all of Baja, and quite an oasis with its spring-fed palms, fig trees, and garden. The *vigilante* gave us a tour. We were his first and only visitors that day, perhaps that week. We stayed too late to go home that evening so we set up camp under a tattered *palapa* on the mission grounds. We cooked pasta and then played with the children wearing tattered clothes who lived in the house. They smiled all the

egories: cardiopulmonary arrest, bleeding, and spinal injuries. The treatment for these requires formal training beyond the scope of this book. Cardiopulmonary arrest, which is the loss of breathing and a pulse, requires CPR. Major bleeding requires pressure, ice, elevation, and other techniques to stop it. A head or spinal injury requires immediate immobilization to prevent further injury and permanent paralysis. You should learn these advanced techniques by taking a wilderness first-aid or Wilderness First Responder course. These may be available at mountaineering clubs or community colleges.

while as we practiced our Spanish and were fascinated by our electronic English-Spanish translator. When we dug out Oreos for desert, the children twisted them apart and licked the frosting.

Two hours for 30 miles, Baja California, Mexico

CANYONEERING SKILLS

- **CANYON HIKING**
- **ADVANCED SKILLS: BOULDERING AND SWIMMING**
- **TECHNICAL SKILLS: RAPELLING AND ASCENDING**
- **CANYON BIKING**
- **CANYON PADDLING**

7

CANYON HIKING

Canyon hiking and basic canyoneering are a great introduction to canyon country. You can start with simple hikes through well-known canyons, then slowly work up to more difficult trips. Canyon hiking employs many basic hiking skills with which you may already be familiar. This chapter reviews those basic skills needed for entry-level canyoneering.

GETTING THERE

Driving to a canyon trailhead is usually straightforward. However, some canyon areas may be down unimproved dirt or gravel roads. Follow an updated road map that includes dirt and gravel roads. Note that many national park or Forest Service maps include logging or mining roads but they may not be up to date.

Make sure your car is in excellent running condition. Get regular tune-ups and safety checks. Before you leave for your trip, make sure to check your vehicle's oil level and tire pressure. Make sure you have the safety and repair tools for your vehicle as noted in Table 1 (pages 34–35). Learn how to perform basic repairs, especially changing a flat tire.

Always use caution when driving on back roads. Drive slowly and watch for animals darting across the road, especially at night. Cattle and deer can surprise you and wipe out your car and themselves. When in doubt, drive slowly and conservatively. If you are unsure if your car is able to pass over a rough road, walk the road first and have someone direct you around large rocks or ruts that may damage the undercarriage.

Watch for sand, mud, and snow. Learn how to drive in these elements and learn how to get unstuck. For example, lowering your tire pressure in sand may improve traction, steady the ride, and lesson the risk for tire puncture; however, you need a pump to re-inflate your tires when back on the highway. Rain or snow can make clay-based soil very slippery, even if you have a four-wheel-drive vehicle. The road conditions can change overnight, especially with the onset of evening thunderstorms.

Previous pages:
Searching for
and exit, Moab,
Utah

If you can, carpool to the canyon and plan a loop hike to save a car shuttle. If you are doing a one-way descent or ascent, consider a hike back on the rim or use a mountain bike. Drop your bike off at the canyon exit and lock it to a tree. When you exit the canyon, whoever has the most energy can ride back to the car.

HIKING BASICS

Canyon hiking is similar to trail hiking but involves getting over and around more obstacles, which may include streambeds, pools of water, waterfalls, logs, and rocks.

Make sure your pack fits well and that you stay balanced when you walk. When hiking, keep your head up. You will need to look down frequently to see where to step, especially when hiking in rocky streambeds. But pay attention to what is ahead and to your pack, especially if you have gear tied on the outside, as it can catch branches. Try to avoid snapping a branch at the person behind you.

Walk on a level canyon floor or trail on a stream bank if you can. Hiking on river rock or logs can be difficult, especially if you are carrying a heavy pack. Use caution and hike methodically and deliberately. Make sure you set your foot down on stable ground. Avoid hopping from boulder to boulder. Watch for loose rocks or logs.

If you are in a streambed with running water, you may stay dry by hiking along the bank. However this can add distance to your hike, especially if the stream meanders significantly. You may be able to take a direct route up a stream if you have wading shoes and socks. However, doing so does mean you will expend more energy and go more slowly than you would on a trail, grass, or rocks on the bank, especially if there are a lot of slippery rocks or soft sand in the stream.

When hiking, watch for major landmarks, especially side streams and rock outcrops that you can identify on the map. Frequently look behind you, particularly when you pass a known exit from the canyon or an access trail to the rim. Note these points on your map.

Remember to use low-impact hiking techniques as noted in Chapter 1, and observe basic trail etiquette. If you meet someone hiking the opposite direction on a narrow trail, the person hiking downhill usually steps aside. This way the person hiking uphill can pass without breaking his or her hiking rhythm. If you need to stop, try to get off the trail to let others hike by without stopping. If you are hiking around someone, let them know you are passing. To avoid a branch in your face, don't follow too close.

Pace

One often overlooked technique of hiking is keeping a good rhythm. Doing so will help you use energy efficiently to maximize your comfort, endurance, and fun. The basic goal is to find a hiking pace that you and your group feel comfortable with, and to keep it steady. Hike too fast and you may make errors, become tired, or get injured; hike too slowly and you won't make progress.

Keeping a steady pace means keeping a steady rate of energy expenditure. On trails, firm sand, or slickrock, you will increase your gait. Up hills, on rocky surfaces, or in water, you will slow your steps. The trick is to adjust your rate to keep the same level of energy output.

Regular rest breaks can also be important for maximizing your hiking efficiency. Find out what works as good intervals with your group, taking into account physical conditioning of members, difficulty of the hike, weight of your packs, the availability of water and shade, and other factors. Plan to rest at regular intervals, say every forty-five to sixty minutes or so. Remember, when taking a break in mild or cool weather, you will chill quickly, so put on a fleece hat or jacket before you get cold.

Fording
a stream,
Columbia
Gorge, Oregon

Good communication is paramount when it comes to setting pace and taking rest breaks. Make sure everyone understands and agrees with how the group plans to proceed. Don't drag someone into the ground on the first day of a three-day trip. Recognize if a member of your party is uncomfortable even if he or she does not speak up; pay attention to important body language. Also, don't be so locked into a schedule or pace that you pass up a worthwhile side attraction like a waterfall or spring. Take some time and enjoy the trip.

Breaking Trail

Breaking trail through thick brush, snow, or mud can be the most difficult part of the hike. Take the shortest route whenever possible. Keep in mind that breaking trail will take you much longer than hiking on an established trail or the canyon floor. Hiking an established trail probably takes less time than breaking a new one that is a third the distance. When breaking a new trail, consider the impact to plants and animals. You may be disturbing the local ecosystem by bumbling through fragile canyon lands.

When breaking trail, look for boulders or downed trees to hike on. When you push or pull through bushes, watch that they don't snap back on your partner behind you. Communicate with your partners, especially when you find a deep mud hole to avoid or a good rock to step on. When hiking, give people ahead of you enough space so as not to crowd them, but at the same time stick together as a group. A good rule is never to let people out of sight or out of talking distance.

Cairn

Using Poles

If you are hiking anything beyond a flat, well-trodden trail, poles are invaluable for hiking efficiently and using as protection. You can use them for support to help you save energy or to catch yourself in a stumble. Use poles to probe for rocks and to give you balance in rough terrain or water. Watch the tips, as they can get caught in crevices and holes. Use small baskets or make sure to remove larger ones so they don't get caught on brush. Use a wrist strap for extra support and to keep from dropping your pole, especially in water.

When hiking with poles, use a light comfortable grip and plant them solidly in the ground. Swing the poles forward with your natural hiking rhythm so you don't lose too much energy. Adjust the pole length as needed: shorter for uphills, longer for downhills. For long traverses, shorten the uphill pole and lengthen the downhill one.

Clothing

See Chapter 3 for a detailed discussion of clothing requirements. Generally, you need gear for either cold or hot and sunny weather.

Remember to balance internal heat and body moisture from exercise with the cold, wet weather outside by using layers and ventilation. In hot weather cover up with long-sleeved, long-legged, light-colored clothing and a hat.

Sunscreen

Sunscreen of at least 20 SPF is essential on all exposed skin. Keep in mind that a white cotton T-shirt is 8 to 10 SPF. You need to reapply sunscreen throughout the day, as it wears off with sweat and swimming.

Food and Water

See Chapter 2 for a detailed discussion of food and water requirements. Keep your water bottle handy and drink often. Never pass up a water source in the desert with empty bottles. Always filter or purify your water first. Eat at regular intervals and include high-quality, high-calorie foods in your diet.

LAND OBSTACLES

Numerous obstacles are encountered on any canyoneering trip. The most common land obstacles are highlighted here, but keep in mind that there you might encounter some obstacles not discussed in the following pages.

Uneven Terrain

Uneven terrain is the norm for canyoneering. Boulders, rocks, log jams, and water-falls all pose particular problems for negotiating. The basics of hiking always hold true: stay balanced, watch your step, stay efficient with your energy expenditure. Be cautious of uneven ground, especially when you are carrying a heavy pack. Use trekking poles for support.

On declines, make sure you can see where you are going. Look for ledges, trees, or depressions to step on. Be careful of jamming your toe or jolting your knees on downhill sections. Tighten your boots, bend your knees, and place your feet lightly to minimize impact.

Never jump from a drop-off. Even when you step down, be careful, as mud, ice, or water can surprise you and can be difficult to see. In some cases you may be able to slide down a steep section. But use caution to control your speed and make sure you have a soft, flat landing. Use a spotter for assistance. Take off your pack and hand it down if you need to.

For inclines and traverses, stay on your feet, use poles for strength and balance, and try to avoid bending at the waist.

Narrows

When negotiating narrows, the sections of canyon that get very tight, you may have to twist sideways or even take off your pack to get through. Keep in mind that you should know what is on the other side of the narrows and be able to reverse your route if there is no exit on the other side. In some cases, turning around may be very difficult, and you may need to hike backwards to get out.

If the side walls come to a V shape at the floor of a slot canyon, you should try to hike with one foot on either side of the V if possible. With feet flat in the V, your feet tend to twist and this makes walking difficult.

Chockstones and Log Jams

Be careful of chockstones and log jams. They can be precariously perched and may crash when you step on them. Climb up and over gently if there is no way to get around them. Be prepared to step off if an object begins to move. Always hike up and over one at a time and use a spotter.

Family of turtles sunbathing

Talus and Scree

Talus is loose rock, and scree is loose rock that is situated on a slope. Hiking up or down talus and scree can be difficult; the rocks can slide or your feet can get stuck between rocks. Look for trails that zigzag up talus slopes. Always watch for unstable rocks and boulders. Stick to the sides of the slope where vegetation may add stability and provide handholds. If there is potential for a rock slide, find a route around the talus or scree. Hike up one at a time and space out. Be prepared to yell "rock" if you kick down a stone. Avoid hiking directly below or above another person.

WATER OBSTACLES

Water can be a hazard in canyoneering. Water obstacles come in many forms. You will likely encounter streams, pools, and waterfalls blocking your path. Before heading into a wet canyon, familiarize yourself with the major water obstacles and learn how to swim.

Streams

When you are travelling in areas where there are streams, make sure you are prepared for water hiking and have proper footwear. Always wear shoes to give traction, warmth, and protection from rocks and sticks. Step slowly with a wide stance to maximize balance and control. Carefully place one foot on a rock to test its stability before putting your full weight on it. Rocks in the water can be extremely slippery, so use caution when hiking. Trekking poles are a must when you plan to do a lot of water hiking. Poles help with balance and support, and can be used to probe for loose stones, pools, or logs. Also try to avoid getting your foot stuck; it can get trapped in a hole or between rocks. Logs and trees may be floating or poorly supported, so use caution when crossing streams on downed trees. Be careful of bounding between dry rocks in stream. You can slip and fall into water.

As discussed above, you may need to choose between hiking from bank to bank, avoiding the water, or hiking directly down the streambed. It will be a trade-off between time and distance. Keep in mind that streams are of two basic types: wide, slow, and shallow or fast, narrow, and deep. Trails on the bank usually meander with the stream, but in most cases it is faster to walk on the bank than in water. Hiking on stream banks can trample fragile habitat, so use caution and disturb the ecosystem as little as possible.

Pools

A good basic rule is to never jump into a pool. It can be difficult to gauge its depth, and submerged rocks can be just under the surface. Probe the pool with a pole or your feet to check for depth, hidden rocks, and submerged logs. Even after probing for rocks, you can still inadvertently jump on a submerged boulder, so use extreme caution.

To cross a pool, wade into it. You can use your pack for flotation if the gear is stowed in drybags. Some people use inner tubes or even inflatable rubber rafts. In a pinch, you can toss in your pack, then straddle your pack.

Keep in mind you may have to climb out of a pool at the other end. If you can't see dry land at the other end

Trail's end,
Escalante, Utah

of the pool, one person should scout with the other waiting. This way you can always reverse the route if needed. Keep a throw rope handy and use a spotter if necessary.

Waterfalls and Dryfalls

Waterfalls and dryfalls, those with seasonal water, are plentiful in many canyons. In basic canyon hikes, you will be able to hike around or climb down them easily. If the falls are dry, hike up or down them carefully, watching for loose rocks or dirt. Use large ledges and trees if they are available.

If water is running down the falls, stay to one side or another to avoid the water. In many cases a trail may skirt the bank on one side or another. Be cautious of the hydraulic force of water; it can suck you down into the water or over rocks. At the bottom of the falls there may be a plunge pool or body of water of unknown depth. Larger plunge pools are often called punchbowls.

CAMPING

Wilderness camping is the topic of many other books and is too complex a subject to cover here. By the time you get to overnight trips in canyons, you should be well versed in wilderness camping in several types of terrain, including forests, mountains, and deserts. When camping you will need the equipment described in Chapter 3 in addition to your day gear. It will help to keep in mind the basic principles of canyon camping outlined below.

Making friends at camp, Baja California, Mexico

- Search for a campsite when you still have daylight and enough time to set up camp. If you are in a high use area, camp in designated sites. Remember you may need an overnight permit. If you are in wilderness with no designated sites, choose a spot keeping in mind low-impact guidelines discussed in Chapter 1.

- Good campsites are usually flat, sheltered areas on sand or dirt. Avoid fragile moss, macrobiotic soil, or plants. Try to camp near a water source but keep your camp 200 feet from water. If you camp near water, make sure you are well above the high water mark and away from flash flood danger. You may have to hike a bit to find a suitable camp, so start looking early in the afternoon.

- Change into dry, warm camp clothes. This is important, as you can cool down fast after a day of hiking. Put on dry socks, camp shoes, and fleece pants and jacket in cold weather.

- When you set up your tent, find a flat spot but don't alter the land. If you absolutely must move a rock or log, replace it the next morning. Make sure your tent is secured by tent stakes and use your rain fly for extra warmth and protection from rain or snow.

SOLO

At the truck, I loaded my fanny pack, including in it nothing but the bare essentials: water, food, a windbreaker, a tiny emergency kit, a camera. I ran a quick mile up the trail to get my blood flowing. At the waterfall and swimming hole I saw no one. I quickly scrambled up the Class-5 rock alongside the falls. I'd done it several times and knew the route well.

Mist from the waterfall sprayed my face and moistened the rock. It was slick enough that I had to pay attention. Up top, I traversed into the canyon, climbed a log jam, waded through the plunge pools, and scrambled up another rock face, this one slick with moss and with water seeping out of a crack.

Deeper in, the trees were bigger and the canyon widened. I hiked on a sloped canyon wall for a mile. Here there was talus, grass, more talus, and a thick grove of scrub oak. Back to the water, the streambed was too narrow to stay dry. I hiked on slick boulders in ankle-deep water, occasionally catching my foot, stumbling. Although soaked, my feet were toasty in amphibious boots and socks designed for such a place.

After an hour I stopped, drank, and ate. Then, the second hour, I got into something of a rhythm, one about as good as you can get when hiking up a streambed. I passed more logs, more water pockets, a deep plunge pool, a few more water-

- You will probably need to procure water for cooking and to replenish your day's supply. Filter, boil, or use iodine tablets, as discussed in Chapter 2. Try to establish one area for cooking to minimize impact. If water is scarce, don't pass up a source to fill all your containers. Another option is to cook your meal in the afternoon at a water source, fill all your water bottles, then hike to a dry camp. When you need to bathe, remember to get water, take it 200 feet from the source, and then wash. Use biodegradable camp soap sparingly.

- When you are ready to cook, use a stove. Don't cook in the tent, as you can get carbon monoxide poisoning or burn down the tent. Use a vestibule that is well ventilated if the weather is wet. A hot meal is nice in the evening or morning but not a necessity. Some ultralight canyoneers don't bring a stove and use meals that don't need to be cooked, like sandwiches.

- Before you head to bed, make sure food is stored properly. In bear country, food should be stored in bear-proof containers stored on bear poles or in trees. This is also a good way to keep food away from rodents, insects, and other small animals. Stow all your gear in your pack in your tent or vestibule if possible.

falls, a mud bank. At last, the dead end. A box canyon. The waterfall was sheer. I might have been able to get around it by hiking back a mile then gaining the rim high on a sloped wall. I probably could have climbed the waterfall with a rope and partner. But that would be another day. Now it was time to head back down.

Far up a side canyon, Columbia Gorge, Washington

8

ADVANCED SKILLS: BOULDERING AND SWIMMING

Eventually, every canyoneer will graduate to more advanced trips. Beyond hiking and wading, this usually involves bouldering, swimming, deep-water wading, stream crossing, and negotiating more difficult obstacles than encountered on more basic trips.

The basic hiking and wilderness travel skills outlined in Chapter 7 serve as the foundation for advanced techniques. There is no clear line between basic and advanced canyoneering, although the proposed classification system in Appendix 1 is a rough delineation. Water is usually deeper and swifter. Streams often take up the entire canyon floor, so extended wading is common. Chockstones and log jams are larger and more difficult to climb up and over; you may not be able to hike around them. Narrows are tighter, sometimes extending for several hundred feet, and may have water, mud, or rocks on the canyon floor below. Scouting and routefinding are usually more difficult and time-consuming. Campsites are much more difficult to find, especially those out of the flash flood zone. Evacuation and backtracking is much more difficult.

It is important to gradually work up to advanced canyoneering. Learn skills well from a climbing, swimming, mountaineering, and canyoneering course. Practice these techniques often at your local swimming pool, rock gym, or crag. Appendix 2 lists several good books that highlight these advanced skills in detail.

Again, you should always be prepared to reverse your route, continue through the canyon, or exit to the rim with the skills you possess. Keep that in mind when planning and traveling through canyons that require more advanced skills.

BOULDERING BASICS

Bouldering is rock climbing that is low to the ground, so there is little or no risk of a fall. A fall is usually just a step down. Bouldering can be easy scrambling or may require more difficult moves involving skilled rock climbing maneuvers. Some basic bouldering techniques can help you negotiate small down climbs, narrows, or pools.

It is important to know some basic concepts. In general, it is best to keep three of your four extremities on the rock and to move one arm or leg at a time. This maximizes your stability. Since your legs are stronger, try to put most of your weight on your footholds and avoid hanging on the rock with your hands. Distribute your weight evenly from right to left to keep balance. When you reach or step for a hold, make your movements smooth and deliberate; move fluidly, conserving both time and energy. Always test the hold, as loose rocks and logs can come tumbling down.

Climbing down is usually more difficult than climbing up. It is difficult to see where you are going when down climbing, and many maneuvers are more difficult to execute. You can down climb facing out if the rock is at a low angle and there are plenty of ledges and hand- and footholds. However, for more difficult exposed sections, climb facing the rock to have more stability and maximize your strength. When facing the rock, it can be harder to see, so use caution when placing your feet and hands. Sometimes you may need to lean out and look for your footholds, then lean back in to step down.

A common hazard in canyon bouldering is getting rim-rocked. This occurs when you climb up a boulder or ledge and cannot execute the more difficult climb down. Don't get too high or far from your starting point without making sure you can climb down. Scout a second exit; you may be able to easily climb down another area. Sometimes a longer, roundabout route is much easier to go up and down than the direct route.

Climbing Maneuvers

Certain maneuvers are designed to help you climb efficiently using strength and finesse. Some moves are commonsensical, and most people who scramble up a large rock or boulder probably employ these techniques already.

When using hand holds, try to grab rock nubbins that are large and stable. Always test your holds to make sure they are not loose. Avoid gripping too hard, as your hands will fatigue quickly, and then pull down, not out. Sometimes you will have to pull yourself up using the weight of your arms. But rely on arms for balance and legs for strength as much as possible.

When using footholds, try not to step higher than the height of your knee and don't fully extend your leg. Keep your legs spread apart—more than shoulder-width but not spread-eagled—for stability. Most of the time you will use an edge, smear, or jam technique. The following list briefly describes some basic climbing maneuvers.

- **Edging** involves stepping on a ledge using the side of your shoe. You use the ball and inside edge of your foot to maximize stability and strength.
- Sometimes you can use the entire bottom of your shoe flat on the rock. Using this technique is called **smearing.** A smear relies on the friction of the shoe sole to keep you on the rock. This works well for dry, low-angled, smooth rock, especially slickrock or bedrock. It may be the only way you can climb up a flat section that

Left:
Down climbing the final waterfall, Columbia Gorge, Washington

Right:
Stemming up the slot, Robbers Roost, Utah

doesn't have any footholds. If the rock is wet or you wear shoes with hard rubber soles, this may be much more difficult. When using a smear, place your feet flat on the rock and stand up, putting all of your weight on your feet. Center your weight over your feet and walk up or down slowly, keeping as much of the shoe sole in contact with the rock as possible. Avoid bending over and putting your hands on the rock; this takes weight off your feet and dramatically decreases your feet's ability to stick.

■ When you wedge your toes or foot in a crack you use a technique called **jamming.** This can be painful but sometimes it is the quickest and safest foothold. The basic maneuver is to place your boot toe or foot in a crack and put weight on your foot. Gently weight your foot and when you are ready to move on, pull your foot slowly out of the crack.

■ A **mantel** can be particularly useful when getting on or off a small ledge. Place your hands flat on a ledge that is about chest high. Then push down with your hands to lift your body up onto the ledge and swing your legs up. You can do the reverse to get off a ledge by weighting your hands, then slowly lowering yourself.

■ **Counterforce** is a technique that uses pressure in opposite directions that can be used when no easy hand- or footholds, such as rocks, are available. A **stem** or **chimney** technique uses feet or hands against opposing walls in a slot canyon. It can be useful when the walls are close enough together to reach with arms or legs.

It can be done using one of several variations: move one arm and one leg on each side to stay upright; face down with hands on one side and feet on the other; or face up with feet on one side and back on the other using hands for balance and support. This is an easy and quick way to bypass walking though water, rocky ground, mud, or a slot that narrows to a V. The basic idea is to use hands and feet in opposition to give you enough friction to stay on the rock or canyon walls just above the slot's floor.

Spotter

Finally, when bouldering or climbing up and over any rock, ledge, or log that has any risk of fall, even a small one, use a spotter to stand by to slow or prevent a fall. The spotter should usually keep a wide stable stance and be on safe ground. Sometimes just a helping hand or extra support is all that is needed. Other times the spotter may need to break the fall. Again, with bouldering, one is not very high off the ground so the risks of falling are minimized.

Hand Lines for Climbing

For short down climbs or to ford streams and pools, a hand line, which is usually quick to set up, gives an extra margin of safety. The easiest use of a hand line with large parties is having a person at each end hold it in a stable stance—feet wide apart and braced against a tree or boulder. However, in most circumstances you will tie it to a tree or boulder for a better anchor.

For down climbing, wrap webbing or rope around a tree or boulder using the midpoint of the webbing against the tree and the two ends in your hand. Then climb down using the webbing for assistance and support. When you get down, pull one end of the webbing to retrieve it.

In some cases you can tie a hand line between two trees or boulders. This can be useful for traverses. The first and last people to cross—usually the two with the most skill—will need to cross without the rope tied at one end. Another option, if you have a long rope, is to tie it in a big loop between two trees. When the last person climbs across, he or she unties the knot and recoils the rope.

WADING AND SWIMMING

Wading is common in all levels of canyon trips, from canyon hikes to technical canyoneering. Always have good equipment, especially shoes and socks designed for prolonged water hiking. If you are hiking any advanced canyons, you should know how to swim. Take swim lessons to learn the front crawl and survival float. Anytime you are in pools or crossing streams, you should consider using a hand line as described below for extra support and safety. Always go one at a time and use a spotter with a throw rope.

Waterfalls and Plunge Pools

If you need to cross a plunge pool, you may need to swim. However, if you are wearing a wet suit and carrying a pack on your front, doing the crawl can be difficult. Often, floating on top of your pack filled with drybags will get you across just as quickly as swimming and requires less energy. Float with your chest on the pack facing the water and kick. Another option is to wear the pack on you back while facing up. However, it is more difficult in this position to jettison the pack quickly, which you may need to do

Punchbowl, Columbia Gorge, Oregon

when you encounter rushing water. Floating on your pack can be particularly helpful if you need to search for an exit on the bank of the pool or stream.

Avoid the hydraulic forces of waterfalls at all times, especially in deep plunge pools and narrows. Anytime the water is more than just a trickle, the force of waterfalls can suck you under and pin you beneath the water. If by accident you get caught in a waterfall and it drags you under, dive to the bottom of the pool and push off to one side or another. This will help you escape the forces of the waterfall and surface in calmer water.

Streams

Be cautious of swift water and water that is more than knee-deep. Submerged holes, rocks, logs, vegetation, and debris can surprise you, trip you up, and pin you under. Always use extreme caution in water. Use trekking poles or a stick for extra support; probe depth of water and for hidden obstacles. Rocks underwater will be super slick, and those just on the surface can be also. Avoid bounding from boulder to boulder as a fall can be disastrous. Rather, step deliberately on stable ground each time.

If you are wading in swift water, you may need to face upstream so the pressure of water doesn't knock you down.

If you need to swim across a stream that is too deep to wade, don't swim directly across the river, as you will be fighting the current. Find an area of still water, if possible below rapids and eddies; swim at a 45-degree angle downstream to end on the opposite bank a bit further down river. This will help you use the current to your advantage when crossing.

Always take off your pack or loosen the straps. If you fall, toss your pack or use it as protection against a fall. If you are wearing it in swift water and you fall, it could pin you under.

If you start floating away in swift water, try to stand up, keeping your feet apart for balance. Use a stick or pole for stability. If the water is deep and you start drifting away, put your body in a seated position, pointing your hands and feet downstream. Bend your knees and keep your feet out in front of you. This will help you fend off boulders and logs. A pack can act as both flotation and protection: hang on to it but be ready to jettison the pack if it gets caught by a rock or log.

Don't fight the current or the natural tendency of the river to meander to one side or the other. Sometimes you can swim to the nearest bank, especially if it is close. Other times you will need to swim to the bank that the river is naturally pushing you to. Often a stream with take you into one bank with an eddy where you can regroup.

Sometimes, for example when you are on a trip where there are large pools, frequent deep stream crossings, or long stretches of deep wading, having a flotation device may help. Some people use inner tubes to float their packs. When an inner tube is

FIRST DESCENT

The first look into the canyon—the wide expanse that encompasses the entire drainage for the small tributary—was surreal. It was like a fantasy poster, a little Narnia world locked into its own, with no outsiders, no humans. I could see my buddy's truck parked in the arroyo. It was loaded with our extra gear. Our second vehicle was parked miles down canyon.

The drop into the canyon came with many months of anticipation: planning, equipment gathering, gear sorting, map reading, route book reading, weather watching. Our planning and research payed off; we made the drop on a hot, sunny day with not a cloud in the sky.

The drop off the rim put us at the headwaters—but it had been dry for a month so there was no water to be found. The wide basin was more than a mile across. We followed a dusty, dry drainage that slowly, over the course of a few miles, deepened. Occasionally we walked past undisturbed macrobiotic soil, a black-widow web, a lizard scampering across our path, a discarded rattlesnake skin.

After a few miles we stumbled upon the slot, almost into it. The crack seemed to come from nowhere, a 5-foot-wide fissure we could jump across. We hiked on the edge for 50 feet on each side: dead end. Our truck was now a half-day's hike away and there was no guarantee we could find a way up to the plateau. No way to go but down.

We set up a rappel on a chockstone and dropped into the slot. Within minutes, the ambience changed. From hot, monotonous trek, we dropped into a dark, wet, cool subterranean world; a welcome relief from the heat of the desert sun. A few more rappels and we stopped to filter water at a water pocket. The murky and mucky water tasted like frogs.

That evening we found a flat spot on the slickrock, dead in the flash flood zone, and camped. It was hot all night in that slickrock oven. The next morning, we made a few more rappels, crossed more water, and took a soothing bath in a deep plunge pool filled with water from the last rain, which had probably fallen a month before.

The canyon exit was sort of anticlimactic; a few miles after our last rappel we realized there would be no more rappels; we were out of the slot and into the hot sun again.

Disappointed, we started longing for the car. After a few more miles, the tributary joined the main canyon. Our car was a few miles up a hot canyon. We had no more water and were trekking through a dry canyon that was 103 degrees in the shade. We stopped to rest every twenty minutes, never passing up any shelter, whether a lone pinyon or juniper or a cutbank deep enough to protect us from the sun's rays.

On rappel,
central Utah

used to float gear, a rope web is needed to suspend the pack above the water. Keep in mind that inner tubes can be heavy when deflated. A personal flotation device may be important to bring along if you know you have prolonged swimming or deep-water wading.

Hand Lines for Swimming

When your party is crossing streams, a hand line can provide an extra margin of safety. Usually the rope is tied to a tree or boulder on the bank or held by a person on the bank. With more than two people, you can tie the rope to trees on either bank, but the first and last person will cross with the rope tied to only one side.

When crossing, don't tie the rope to your waist, harness, or pack: it can hold you under if you were to fall. Head across the stream at 45 degrees upstream or downstream so you are not fighting the current. If you fall, get up immediately and use the hand line for assistance. Cross one at a time, placing your feet carefully as you walk. For an extra margin of safety, a person on the bank should have an extra rope to throw if the person crossing falls and loses the rope.

Throw Rope

With any water crossing and in some bouldering situations, you should always use a spotter with a throw rope. For water rescue, a throw rope is coiled in a bag. The thrower holds the end and throws the bag, usually from a downstream position. The person needing assistance then can grab the rope.

COMBINING SKILLS FOR ADVANCED CANYONEERING

One of the thrills of canyoneering comes when you put together several advanced skills on one trip. You may down climb a dryfalls, wade across a plunge pool, then scramble back up to the trail. Hiking through narrows, you might encounter water or even a chockstone that you will need to scramble up and over.

On advanced trips you must be ready for varied terrain, weather conditions, and obstacles; you will need to combine skills. Use all your advanced canyoneering skills and keep in mind the safety of all your party members. Use your pack and drybags for flotation; be prepared to use your wet suit and wading shoes; know how to set up hand lines. Tackle difficult problems one at a time and use a spotter. In many canyons you may be able to minimize exposure to dangerous situations by circumventing obstacles.

EXITS FOR ADVANCED CANYONEERING

As mentioned earlier, you should allow for a broad margin of safety, always making sure you have at least one exit from the canyon other than your planned route. For advanced routes, this usually means you should be able to continue through the canyon and be able to backtrack. For all trips, you will reach a point where continuing down your path will be easier and shorter than backtracking, so long as you know you can get through your hike without any obstacles too difficult for you to overcome. Remember, a recent and unexpected log jam, chockstone, landslide, or high water can stop your planned route.

In many canyons, you can backtrack without difficulty. However, when backtracking, keep in mind that the route back may be much more difficult. For example, scrambling up a dryfalls may have been much easier than climbing back down. Similarly, crossing streams may be much easier in one direction; often it's easier to walk upstream.

Another option is a hike to a rim. Sometimes you may have a known trail but you may have to scramble over talus, brush, or dirt. Know that sometimes a potential route to the rim can stop at a dead end. A long hike toward the rim of the canyon can end on a 10-foot blank wall. Similarly, a long traverse on a ledge can be fruitless if you don't know that the trail will eventually lead to the rim. Finally, a hike to the rim may put you a long way from your vehicle, especially if you have to go across another canyon or two to get back.

9

TECHNICAL SKILLS: RAPPELLING AND ASCENDING

Technical canyoneering is a unique and thrilling sport. It usually involves rappelling, ascending, and rock climbing.

This chapter does not provide a complete description of rappelling, ascending, or climbing; rather it is an overview to guide you to further skills. *Mountaineering: The Freedom of the Hills* (see Appendix 2) is an excellent book that introduces these skills. Few courses on canyoneering exist; many rock climbing and mountaineering courses give an excellent introduction to knot-tying, using equipment, setting anchors, climbing, belaying, ascending, and rappelling.

Keep in mind the safety issues, highlighted in earlier chapters, regarding technical canyoneering. Perhaps most important is to always maintain an exit: know you can continue through a canyon with the skills and equipment you have; have a known exit to the rim at all times; or be able to reverse your route. This will be further discussed later in this chapter.

BASIC RAPPELLING TECHNIQUES

Rappelling is basically sliding down a rope using a tie-in and breaking system. It can be dangerous, and common problems associated with the technique include anchor failure, improperly used equipment, and rappellers sliding off the end of the rope. The overview of rappelling provided here is fairly brief. Books such as *Rappelling* by Tom Martin or the *CMC Rappel Manual* (both of which are listed in Appendix 2) will help you learn rappel skills.

Knots

You should learn different types of knots for a variety of applications and practice them often. You don't want to be figuring out how to tie knots in the field, so you should be able to tie them easily, even with cold, wet fingers, stiff wet rope, and in low light. In general, make sure your knots are tight and leave long tails in the rope—at least 2 inches.

Tie the tails again with an overhand knot as a backup against loosening. If you have any question about the integrity of a knot, untie it, inspect the rope, cord, or webbing, then retie the knot. Below are brief descriptions of the most common canyoneering knots.

Lowering gear, southern Utah

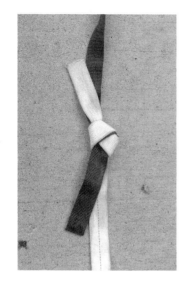  

Left: Water knot

Center: Double fisherman's knot

Right: Figure 8 knot

- The **water knot** is an overhand knot used to tie a piece of webbing into a sling or runner. This can also be used to tie two slings together.

- The **double fisherman's knot,** also called a **grapevine,** is used to tie a piece of cord into a sling or runner. It is used for tying two ropes together and can be used with ropes of different diameter.

- The **figure 8** is the standard knot for tying climbing rope into a harness. It can also be used to tie a loop in the end of a rope or to tie two ropes together.

- The **bowline** is used to tie rope into a loop at one end. It is used, for example, when you need to tie a rope around a tree. It is simple to untie and easy to adjust.

- The **prusik** is a friction knot. It is designed to slide easily when not under tension and tighten when under tension. It is used for ascending a rope. In general, a prusik should be tied with cord smaller than the rope you want to ascend. For example, 6-mm cord makes a good prusik for ascending 9-mm rappel rope.

- The **Klemheist** is a friction knot similar to a prusik that is usually used when webbing is used for ascending larger-diameter rope.

- A **Münter hitch** is a special knot that is tied on a carabiner so it can be used to rappel. This is useful especially when you lose your belay/rappel device and need to use a carabiner as a backup.

Anchors

Anchors provide you with the means to tie your rope so you can lower yourself. They are essential to have when it comes to setting up a rappel station. It is standard to use two anchors in case one fails. Sometimes you may only have one available; if it is bomb-

proof and you are only rappelling a short distance, this may suffice. Use good judgment when inspecting and setting up anchors. If you have any doubts about the integrity of an anchor, use another or consider two backups.

Natural anchors are the most readily available. Depending on your ethics, they may be the only type you are willing to use. Thick trees with deep roots make great anchors. They should be stable, alive, and be at least 6 inches in diameter. Huge rocks also make good anchors. Sometimes a large horn, large boulder, flake, or arch can be used. Make sure the anchor is solid and not loose. A hollow sound or movement when pounding on a rock anchor indicates instability.

Downed logs are often usable, especially if you can wedge them in slots. Make sure they are not rotten and can't budge. Natural chockstones or logs wedged in cracks, crevices, or slots, can be problematic. They make great rappel anchors as long as they

Left: Prusik knot

Right: Münter hitch

Rappel station: Rock arch

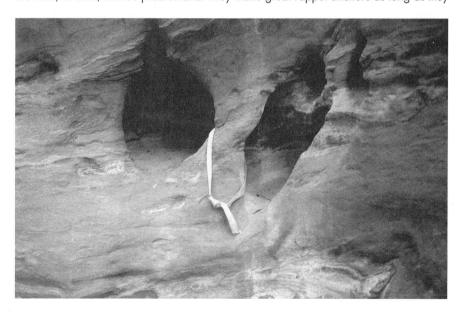

Rappel station:
Log

are wedged tightly with no chance of failing. However, they can be difficult to evaluate, and the weight of a rappeller may be enough to cause a precarious chockstone to break free. With more advanced instruction, you may be able to use a stack of rocks, or bollard, as an anchor.

Artificial anchors come in a variety of forms. Removable anchors include passive chocks and camming devices. If you are using these for anchors, keep in mind that you will not be able to retrieve them after the last person rappels. Not only does this get expensive, it also leaves unsightly hardware in the rock after you leave. Try to leave an artificial anchor only in emergencies.

As mentioned in Chapter 1, permanent protection such as bolts and pitons are controversial. If you are descending or ascending a canyon that has these fixed pieces in place, be careful. Over time they can weaken and loosen. You should always be wary of any piece you have not placed yourself. In those situations where you rely on fixed anchors to descend, you should be prepared to place your own new anchors, should those left by others be unusable. Always use a backup.

Rappel Station

Your rappel station is the entire system you use to fix your rope, and includes anchors, webbing, and rope. To set up your rappel station you will need webbing and rappel rings. Webbing protects your rope from being damaged by an anchor. Rappel rings minimize friction and save both the webbing and the rope from wear and tear.

Always set up a rappel station using new webbing and new rappel rings. Cut away or untie old slings left by previous parties and pack them out. It is not recommended that you use rope directly on an anchor as this can cause significant abrasion to your rope.

Rappel station: Rock bollard

When tying a natural anchor, place a piece of webbing around the anchor, put two rappel rings on the webbing, and tie the webbing into a loop with a water knot. If using a tree, tie the webbing as close to the ground as possible. If using a rock, make sure the webbing won't slide or shift. Pad all sharp edges if you need to.

When using two anchors, keep in mind some general principles. First, always equalize the load, or distribute weight evenly between two anchors. One method to do this is to use one sling for each anchor and attach rappel rings to both. Not only does this equalize the load, it provides separate webbing for each anchor, so there is a backup if one fails. Another way to equalize the load is to use one long sling to connect both anchors. When using one sling for two anchors, you need to put a twist in one side of the webbing, as

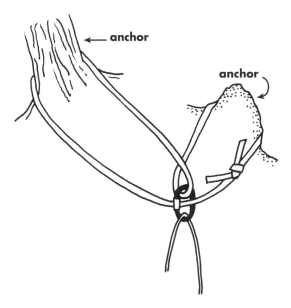

anchor

anchor

fig. 5
Rappel station using one sling. Note twist to "tie" sling into carabiner and prevent sling from sliding out of carabiner if one anchor were to fail.

shown in fig. 5. The twist prevents the rappel rings from sliding off the webbing, should one anchor fail.

A common method of setting up a rappel is to use the webbing in a triangle between two bolts or anchors. However, this can put more stress between the anchors instead of directing the force downwards. A better method is to use one sling for each anchor, as shown in fig. 6. If one sling is used, it should be in a V with a twist in one side, as described in the previous paragraph, in order to equalize the load. Also, a longer sling creates a narrower angle between the two anchors and so lessens the stress on them.

Set up a rappel to minimize rock and dirt abrasion on your rope. Use a rope protector such as a piece of hose or tubular webbing against the rock if there is a sharp ledge. Sometimes you will need to use a long runner to extend your rappel rings past a sharp lip to protect your rope and make retrieval easier.

Once you have your rappel station, pass the rope through the rappel rings. The standard setup is to rappel on one rope that is doubled over. This allows you to pull the rope at the bottom of the rappel. Thread your rope to its midpoint. If your rope is not long enough to reach the ground, tie two ropes together using a double fisherman's knot. Make sure your rope slides smoothly though your rappel rings so retrieving it is easy.

Before you throw the rope over, make sure you tie off the rope near the rappel station so it doesn't slide through the rings when you throw it. To minimize kinks and knots in the rope, be careful how your throw it over. Usually you can throw it in four sections. Gather the rope in four coils, keeping two separate coils on either side of the rappel station. First throw the coiled quarter nearest to the rappel station on one half. When it is down, throw the second coil on that half. Repeat this for the other side.

If there is any possibility of the rope not touching the ground, tie a knot in each free end to prevent you from rappelling off the end of the rope. If you cannot see your ropes touch the ground, be prepared to use two ropes or make a multiple pitch rappel as described below.

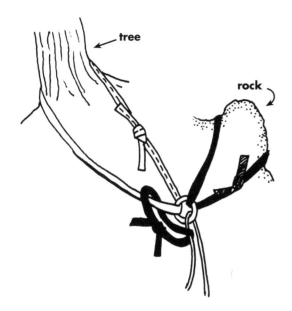

Backing Up Your System

To always assure you have a wide margin of safety, always use a backup system. You can use a backup belay, which is a separate rope tied to the rappeller and belayed by a second person with separate anchors. A backup belay is often referred to as a top rope and can be belayed from the rappel station or from the bottom of the rappel. The safest backup is to put everyone on a backup belay, so no person is at risk; this is a good idea with less experienced rappellers. Belay everyone on a backup rope using separate anchors. The last person can be belayed from below using a top rope.

Another method for backup is for a person already on the ground to stand by and pull the rappel line if needed; this brakes the rappel should the climber lose his or her grip.

A self-belay is a backup break system that gives you a measure of safety if your break fails or you lose grip on your break hand. This is usually a prusik knot placed above your break and tied to your harness. If you lose your grip and the break fails, the prusik should tighten and stop your rappel.

Rappelling

To begin your rappel you need to tie in using your harness with a breaking system. The easiest breaking devices to use are tubes; they are described in detail in Chapter 3. You can tie them to your harness with a retention cord in case you drop your belay/ rappel device.

A figure 8 is perhaps the most versatile type of knot as it can be tied in several ways to change friction. If you use an 8, learn to tie it in such a way that it doesn't need to be taken off the carabiner. You can use an elastic band to keep it from sliding off. Or learn to thread the rope through your 8 first, then unclip it from your harness, finish the tie, then clip it into position. This way, if you drop your 8, it will be attached to the rope at all times.

In case you drop your belay/rappel device, you should have a backup device or

fig. 6
Rappel station using two slings to equalize the load.

learn to use a carabiner with a carabiner wrap or a Münter hitch.

Once your have your rappel device tied and clipped to your harness with a locking carabiner, check and recheck the system. Break with your dominant hand and never let go. Use your other hand for clearing knots or helping yourself over the ledge.

Before rappelling, notify your companions that you are tied in and ready by saying "On rappel." Test the station by applying weight slowly while on a backup belay. Then climb down below the rappel and gently lean out to put your weight on the rope. While on a backup belay, check the anchor again. Slowly feed the rope through the tube or 8, never releasing your break hand.

When rappelling, keep your feet spread at least shoulder-width apart. Keep your knees bent and lean back. Your feet should be about level with your waist and against the rock. Feed the rope slowly and steadily. Watch below you for loose rocks and dirt. Don't bounce or stop suddenly, since this adds extra force to the rope.

When you get to the bottom after making a rappel, make sure your feet are on solid ground and unweight the entire system before you unclip. After you have unclipped, notify your partner by shouting "Off rappel."

Rappelling With a Pack

If your rappels are short, you might be able to wear a small pack. You can make your rappels quickly. However, wearing a pack can throw off your center of gravity. It can pull you out and away from the rope. You might want to consider using a chest harness if you plan to rappel with your pack on your back.

One option is to clip the pack below you so it pulls downward on the rope. Clip a short piece of webbing to the top of your pack and to your harness. After you get on rappel, lower your pack below you. It should hang freely and pull downward. You may need to guide it around bushes or over ledges from time to time.

For many cases, lowering your pack separately works well. This can be time-consuming with large groups and long rappels. You don't want to risk separating from your gear, so be careful when lowering packs.

Pulling the Rope

The last person to rappel should make sure the rope is arranged at the rappel station to minimize kinks and twists. It should be arranged so it can't get stuck in cracks or brush; it should slide smoothly through the anchor.

Once the last person is on the ground, make sure you untie any knots you put in the ends of the rope. Take out all the twists, kinks, and knots. If you are using two ropes tied together, make sure you pull the rope with the knot on your side of the rappel station, otherwise you will pull the rope against a knot that could get stuck in your rappel station.

When you pull, use smooth, even pressure. Just before the rope slides through the rappel station, yell "Rope" so everyone knows it's coming down. If the rope jams when you are pulling it, whip the rope or pull it to one side or another.

On rappel, Utah

ADVANCED RAPPELLING

This is only a brief introduction to rappelling, as there are many variations on technique and more complex skills that you can learn. As your canyoneering skills increase and as you progress to more difficult routes, some advanced rappelling techniques are particularly useful.

Rappelling in Water

Water poses particular difficulties to rappellers, whether in a waterfall, plunge pool, or stream. Above all, use caution. Never rappel in a waterfall, as the hydraulics can suck you down into a deep plunge pool; if you are still tied to the rope after being taken down, it can be impossible to escape.

When you are rappelling near waterfalls, try to start your rappel to the side of the running water. Make sure you will land in water away from the waterfall. If you end up close to the hydraulic forces of the waterfall, unclip while hanging above the water if possible, then push off the wall using the rope to swing to a safer area. If you do get caught in the hydraulics of a plunge pool, dive to the bottom, then push off to one side or the other to escape the suction.

If you must unclip from your rappel in a plunge pool, unscrew your locking carabiner just before it hits the water and make sure the rappel device is attached with a leash. If you are wearing your pack or have it clipped below you, use it for flotation.

Remember, always use caution and exercise good judgment. Many canyoneering accidents occur when people are rappelling into water.

Retrieving Your Webbing

To adopt a truly leave-no-trace ethics style, learn how to retrieve your webbing and rappel rings. This involves using a separate retrieval cord, which is sometimes called a cordlette or reapschnur.

Tie a long runner with a carabiner or rappel ring at each end, as shown in fig. 7. Wrap the runner around an anchor, such as a tree, without tying it. (This differs from the way you usually place a runner, which is looped around the anchor and tied.) Next, run your rope through both carabiners or rappel rings. Tie a retrieval cord to the webbing above one of the carabiners; the cord should be long enough to reach the ground. After the last rappel, pull the rope carefully, making sure you don't twist the webbing above. When the rope is down, pull the retrieval cord to bring down the webbing with the two carabiners or rappel rings.

One-Rope Rappelling

A one-rope rappel is used so one can carry less rope or make a longer-than-expected rappel. This techniques allows you to get down a 150-foot pitch with one 150-foot rappel line instead of using twice the length of rope and doubling it over. However, you need accessory cord the same length as the rope to retrieve the rope. In most cases, you tie

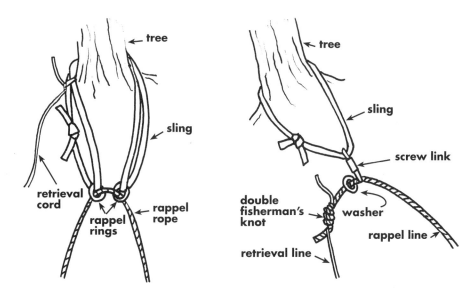

the rope and retrieval line together, with the knot in the rope holding the rope against the anchor, as shown in fig. 8. When rappelling on one line, keep in mind you have much less friction than in a two-rope rappel. Once down, pull the retrieval line to get the rope down.

fig. 7 (left) Rappel station using retrieval cord to remove webbing.

Multiple Pitches

A multiple pitch usually refers to a rappel that is longer than one standard rope length. A standard length of climbing rope is 50 meters, so with two ropes tied together, a 50-meter (150-foot) rappel is usually designated as a single pitch. For multiple-pitch rappels, you have several options. One is to use more that two ropes by tying several together. This involves passing the knot through your belay device while rappelling, which can be complicated and dangerous if not done correctly. Keep in mind that you will need a knot only on one side of the rope or you won't be able to retrieve the rope; this only works if you have one long rope and two shorter ones. A second option is to tie two ropes together and use a combination belay and rappel. Everyone but the last person is belayed down the first portion of the pitch, and then rappels the second portion below the knot. This allows all but the last person to descend without passing the knot. A third option is for a climber to stop part way down and set up another rappel station while on rappel. In some canyons with overhanging or free rappels, this may be difficult.

fig. 8 (right) Rappel station using one-rope technique

ASCENDING

Ascending is another advanced skill. You will likely need to take an advanced climbing course to get instruction in this technique. It can be an essential skill in several

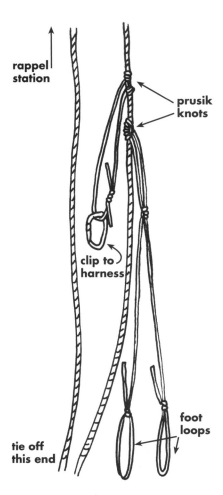

rappel station

prusik knots

clip to harness

foot loops

tie off this end

fig. 9
Ascending with
prusiks

situations: to get ropes unstuck, to reverse your route if you run into impassable sections of canyon, or as an emergency escape to the rim.

When ascending, you will usually be climbing back up one side of the two sides of the doubled-over rope. Tie off one line or have a partner lock it in a belay. If you have backup belay from a second top rope, this will give you added security. Use ascenders or prusik knots, as shown in fig. 9, and use one of several techniques.

A common technique is the inchworm method: clip the top ascender to your harness and the lower ascender to two foot straps on your ascending slings. With one hand on each ascender, your feet should be in the slings tied to the lower ascender and your harness should be tied to the top ascender. Put all the weight on your feet and stand up. This releases the tension on the top ascender while keeping your weight on the bottom one. Then slide the top ascender as high as possible. Once you have slid the top ascender, lock it in position and transfer the weight from your feet to your harness and the top ascender. This releases tension on the lower ascender and allows you to slide it up. Stand up on your feet and repeat the whole process.

This repetitive sliding of ascenders allows you to inch up the rope. It seems tedious but you should be able to ascend 150 feet of rope in fifteen minutes with practice. You will need to keep yourself upright, which can be difficult; a chest harness makes this easier. For safety, tie a loop in the rope below you every few feet and clip it into your harness. This serves as a backup belay in case one or both of your ascenders fail or come off the rope.

Another method that is faster but more difficult and energy-draining is a method used by climbers called jumaring. It involves using a separate foot loop and harness sling for each ascender. The top ascender is clipped to your harness and to a sling for your right foot. The lower ascender is clipped to your harness and to a sling for your left foot. The top ascender is weighted; the lower slides up. Weight is transferred to the lower ascender, and the top ascender is then slid up. By alternating right and left, this method allows for quicker ascent than the inchworm method.

ROCK CLIMBING

Rock climbing is an advanced skill not fully covered here. This involves using passive, removable, and fixed protection for safety, as mentioned in Chapter 3. You will need special gear including a climbing rack of protection pieces and a lead climbing rope. Take a climbing course and practice techniques at your local gym or crag. Appendix 2 lists several good rock climbing books to help you learn techniques.

COMBINING SKILLS IN TECHNICAL CANYONEERING

Part of the fun of canyoneering is combining skills. Combining might mean rappelling into a plunge pool, swimming across the pool, climbing out, then hiking to the next rappel. Not all canyons will require this type of intense combining of skills. However, you should be ready for varied terrain, conditions, and obstacles. You may be surprised that at the bottom of a rappel is a deep pool or that after swimming across a pool or stream you may hit a wall that is difficult to climb. Remember that canyon conditions change from year to year and season to season, sometimes dramatically. Be prepared for the difficult sections and ready for surprises.

Practicing ascending on a backyard tree

EXITS IN TECHNICAL CANYONEERING

As mentioned earlier, the primary measure of safety for any canyoneering trip is that you always know you can continue through the canyon with the skills and equipment you have, have a known exit to the rim at all times, or are able to reverse your route.

In most situations, continuing through the canyon where you have planned campsites

IN THE EARTH

Our group stopped to rest after six short rappels, just before the first big drop—150 feet big. We huddled in the dark cavern in full wet suits, gathering our breath. I tried to eat but the experience was so intense deep in this canyon I became nauseated. Despite the air temperature nearing 100 on the rim, we shivered in the cool slot canyon. The first six rappels—although each only a short 20 feet by climbing standards—were difficult and exhilarating: through rushing waterfalls, over log jams, around blind corners, across slippery slickrock, into deep pools. Standing at the big drop deep inside the earth we gathered our nervous breaths.

We wondered where the deaths had occurred two years before; an eerie tone filled the cavern when we spied a small plaque fixed to the sandstone with expansion bolts. We could still see remnants of the rock pile that they had huddled on for five days. We inspected rescue webbing still in place. We tried to imagine what it was like—raging water, dark, cold, no direct sunlight, hungry, shivering. Later on in the day, we saw shredded fabric from a sleeping bag, an exploded first aid kit, a pack frame. Months after the trip we would pass around old news clips my friend had dug up from the library; we knew first hand what the ordeal entailed. Eleven rappels, half a mile, and twelve hours after leaving the rim road we found camp, barely big enough for our group but with a few trees and some shrubs so it felt like an oasis. The next day would be more of the same, but not knowing exactly what rappels, pools, slots, logjams, and chockstones lay ahead, we still felt a spirit of adventure as we sat there two full days away from civilization.

After cooking, setting up our bivys, laying out clothes to dry, and sorting gear, we built a small fire along the streambed to dry our clothes, quell our shivers, lift our spirits, and perhaps lift theirs, too.

and exits is often the safest and quickest option. However, your way can be blocked by log jams, landslides, high water, and many other obstacles. You need to have a means for exit beyond continuing through your route.

In canyons that require technical skills, retreating the way you came will not be easy, especially if your route is heading down canyon and you need to retreat back up. You have a few options. One is to leave your rappel lines in place until you are assured you can continue to another spot with a known exit. In extreme cases this may mean bringing along several rappel lines and leaving them fixed in place. When you are sure the canyon is passable, you need to ascend to the first rappel and then pull all the lines as you rappel each one again. Another option is to carry a climbing rack to make sure you can climb back up the canyon. In some cases a bolt kit or piton rack may be needed for aid climbing.

If your direction of travel is up canyon, make sure you have a long enough rappel rope so you can retreat by rappelling.

Another option when going up or down canyon is to make an exit to the rim. Again a full rack of climbing hardware, including a bolt kit and piton rack, may be necessary to do this. Having artificial protection allows one to free or aid climb out of a canyon where no exit exists, using the so-called blank wall exit.

In some extreme cases, for example on first descents, canyoneers have been known to hike the entire canyon rim first to find an exit. Finding no exits readily available, they drop a rope or two to be later used as an emergency exit to ascend to the canyon rim.

10

CANYON BIKING

Biking is one of the most popular outdoor sports and can be a great way to enjoy canyons. However, biking poses additional challenges, especially with regards to routefinding, negotiating obstacles, and packing for an overnight trip. Conditions for biking vary widely from canyon to canyon. Many basic, entry-level rides are through dry washes, on well-known trails, or along abandoned roads. These routes are a great place to start, especially in areas that have mountain bike guidebooks and trail maps. As you get better, you may choose more technical rides on single tracks that may require fording streams and negotiating obstacles such as downed logs or boulders.

Slickrock spinning, Moab, Utah (photo © Eric Idhe)

Most everyone knows how to ride a bike. But canyon biking, like mountain biking, is a learned skill that takes a lot of practice, some formal instruction, and experience. Always ride within your skill level. Remember that you can get off your bike and walk over a log or stream if you must. Many accidents occur when riders lose control on a downhill stretch. Always stay balanced and in control. Watch for terrain changes such as slickrock, sand, dirt, gravel, snow, and mud. Keep an eye out for obstacles such as downed logs, streams, or sudden turns in the trail.

Keep in mind that canyon biking is a demanding sport requiring special equipment, technique, and skill. Consider reading a mountain biking book, studying a biking guidebook and map, and taking a course or guided biking tour in canyon country. This chapter serves to provide a general overview for those looking for basic information.

EQUIPMENT

Mountain bike equipment can be difficult to buy with so many brands and accessories on the market today. Table 3 lists basic canyon biking gear. This is a supplement to the general equipment list for canyoneering given in Table 1 (pages 34–35).

Table 3

ADDITIONAL EQUIPMENT FOR BIKERS*

CLOTHING

    Bike helmet

    Hat fitting under helmet

    Gloves, fingerless or full fingered

    Bike shirt

    Padded shorts or tights

    Rain suit

    Cleats or shoes for toe clips

BIKE GEAR

    Headlamp for helmet and/or bike light

    Packs and/or panniers

    Rear and front racks

    Water bottle cages

    Personal repair kit

    Tire irons

    Spare tube, more than one for long trips

    Patch kit

    Hand pump

    Tire gauge

BIKE GEAR, CONT.

Spare nut and bolt for rack

Thorn-proof tube material

GROUP TOOLS

Allen wrench set

Crescent or socket wrench set

Screwdrivers, flat and Phillips

Chain rivet tool

Valve stem tool

Spoke wrench

Chain links

Spokes

Brake cable

Derailer cable

Valve stem

*This table lists items to take on a biking trip in addition to those listed in Table 1 (pages 34–35), General Equipment for Canyoneering. Make sure all repair tools and parts are compatible with your brand of bike and accessories.

When shopping for a bike, try to buy the best you can afford. They range in cost from a few hundred to several thousand dollars. Choose a sturdy frame that fits you well and which can withstand rugged biking in canyons. Make sure you get the right size; the shop where you buy it can probably help you with proper fit. Have the shop personnel adjust the saddle and handlebars also.

Consider getting both front and rear shocks. Some rear shocks don't allow for a rear rack and panniers, which is limiting for overnight trips. Climbing handlebars, a custom saddle or pedals, and cleats are also important accessories that help improve riding efficiency. Tires should have beefy tread, and gears should be for trail riding. Don't forget that you will need a few cages for water bottles, a backpack hydration system, or both.

Clothing for biking consists of padded shorts or tights plus a bike top that has easy-to-reach pockets on the back panel. Rain gear designed for biking is also helpful, as the top usually has a long tail to cover your butt while peddling, and the pants have tight cuffs so they don't catch in the chain or peddles.

Remember never to skimp on safety equipment: it is designed to save your life and prevent permanent disability. An approved helmet is mandatory to protect your head. Get one that fits snugly but comfortably. It should ventilate well and be adjustable in case you need to wear a hat underneath. Gloves are important, as when you fall you usually stretch out your arms to catch yourself. Sticks, rocks, and logs can cut your hands easily when you don't have the protection of gloves. Eyewear is important to block damaging rays from the sun as well as for keeping tree limbs or dirt out of your eyes. Other safety gear available for mountain and off-road biking includes elbow pads, knee pads, and trunk protectors.

Take some repair tools; see Table 3 (pages 129–130). Everyone in the group should at a minimum carry a pump, spare tube, patch kit, and tire irons to fix a flat tire. A multipurpose tool is also nice to have. It is helpful for long trips to carry spare chain links, spokes, and cables for the derailiers and brakes. Make sure these parts are compatible with your bike, as they come in a wide variety of brands. Even spare tubes come in two types, presta and schrader. If you have repair parts that won't fit you bike, they are useless. Sometimes this means that everyone needs to carry his or her own spare parts.

PACKING

Packing for a day ride can be easy: you will need food, water, and the essentials listed in Table 1 (pages 34–35), which can fit into a fanny pack or under-the-saddle pack. Make sure you have plenty of water either in water bottles or a backpack hydration system.

Packing for an extended ride, especially an overnight, is more challenging. You have several options when it comes to getting all your camping gear on your bike. Since space is limited on a bike, you can go ultralight and pack only the bare minimum. For example, some bikers take a bivy sack instead of a tent or sleeping bag on midsummer overnight rides.

Panniers, or bike packs, are designed to fit over a front or rear rack. These are the standard pack bags for biking. You can also find bags that strap to the top of the rear rack. Some riders tie a sleeping bag and pad on the top of the rear rack. A small backpack distributes the load off your bike onto your body, which makes peddling easier; oftentimes this is easier on you than loading down your bike racks with gear. You can also get small packs for the handlebars, for underneath the saddle, and for the tube triangle near the seat post.

Another choice is a sag wagon, a support vehicle that carries all your gear. Usually everyone in the group takes turns driving the sag wagon, unless you can get a nonbiker to come along as driver. Bike trailers are also becoming more popular.

Yet another option is to drive the route ahead of time and leave a cache of food, water, and gear along the way.

Keep in mind that, loaded down with gear, your bike will handle differently. It will be slow to start and stop, difficult to steer, and heavy especially when riding technical terrain. Anywhere you can store gear helps, but the more weight you have on your bike, the less maneuverable and more difficult the peddling will be. When on your first overnight bike trip, it may be easier to establish a base camp at your vehicle and make day rides until you gain more experience.

PLANNING A RIDE

When planning your bike trip, use the routefinding, weather reading, and other skills outlined in Part I of this book. Remember canyon biking has many obstacles, including rough terrain and bodies of water that you must get around. Some minor obstacles for hikers may be more pronounced for bikers, especially if you must hike up and over a chockstone or down a small waterfall. Some trails are not suitable for bikes, and bikers should use caution if using a guidebook and map designed with hikers in mind. Maps may not show simple obstacles that will stop a bike.

Make sure you allow plenty of time for your ride. Single tracks with mud and sand are much slower than a wide, firm, abandoned road. Similarly, if your bike is heavy with overnight gear, you won't make as much progress as you do on a day ride. Whatever the case, allow time for breakdowns, rest stops, viewpoints, or side trips to a swimming hole.

Bike routes are usually loops or down-and-backs. Loop hikes are great, as you get new scenery and trail the whole route as well as a sense of accomplishment. However, if you have a problem, you may be away from your car, and if you run into an obstacle toward the end of the trip, you may have to backtrack most of your route. Down-and-backs are nice because you can turn around at any time, such as when you get tired or if the weather turns bad. A steady climb up and then a cruise downhill is great on a down-and-back.

Also make sure you are riding in an area where bikes are allowed. Many wilderness areas, national parks, and national forests have limitations as to where you can take bikes. Learn the rules before you go.

LOW-IMPACT BIKING

Canyon biking, like other forms of wilderness recreation, can damage the ecosystem. Always be careful and follow the low-impact guidelines outlined in Chapter 1. In general, bikers should always stay on a designated road or trail. Stay on firm dirt or bedrock. Riding in mud, sand, or loose soil can cause significant damage to the trail or surroundings. Always avoid riding on fragile macrobiotic soil, moss, meadows, or other delicate turf. Keep your speed controlled on downhill portions of trail and don't skid, as this tears up the trail.

Try to follow trail etiquette as well. Always yield to hikers and pack animals.

Facing page: Redrock descent, southern Utah (photo © Eric Idhe)

SLICKROCK SERENADE

Quick start, click in, spinning. Gravel turns to dirt turns to slickrock, redrock. So smooth. Up the first grade the soft rubber knobbies stick to the sandstone, then when it gets steeper they slip once, twice, three times. Still cranking, I stand up in the saddle, put more weight on the rear tire, get better leverage. I crest the hill. At the top the slickrock goes on for miles.

This well-ridden path is marked by black rubber streaks, the sign of the canyon biking tribe. It's difficult to get lost on this ride. Round and round, everyone follows the same path in a big loop to get back to where they started. Strange. Other routes are more scenic, more technical, more physically demanding, more mentally challenging, more solitary, more fun. None is so popular.

Now I'm cruising the flats, an occasional flat spot of sand slows my pace or a dry water pocket gives me a jolt. Shift down. Slow, negotiate a fallen chockstone, long since broken, cracked in two from a hundred-foot fall a thousand years ago. Keep pushing, keep spinning. I've seen the redrock arches and bridges of this familiar territory many times but still it's enchanting. A big playground for adults who act like kids. Zipping along, the grade steepens. Thighs burn, back arches; legs, ankles, feet, arms, and hands transfer energy into the handlebars and peddles, into the frame and down to the tires that grip the hot sandpaper rock. Generating more heat, I'm sweating, dripping. My water is going fast.

I'm drinking more and more, quenching my parched throat. Occasionally I feel like I need to spit but I don't want to loose precious moisture so I swallow the dry mucous, and keep peddling.

11

CANYON PADDLING

Paddling craft come in many forms. Canoes, kayaks, and rafts are the most popular, although all types of boats and even inner tubes drift canyon streams. Canyon paddling, no matter what type of boat is used, poses additional challenges. Paddling through canyons varies from easy streams to technical descents requiring running rapids or portage. The beauty of canyon paddling is that you have plenty of room for gear, can enjoy the canyon from the stream, and have the thrill of running a river.

Like biking, paddling is a demanding, technical sport requiring specialized equipment, technique, experience, and skill. It is the subject of many books. When starting out, choose calm waters and consider taking a guided trip to learn the basics. Always stay balanced in your boat and watch for obstacles. Keep your paddling motion smooth and rhythmic. Begin your paddling education with a book on canoeing, kayaking, or rafting, then consider taking a formal course or guided tour in canyon country. The discussion provided in this chapter is not nearly detailed enough to instruct one on canyon paddling. Rather, it serves to provide a general overview for those looking for basic information.

EQUIPMENT

Boats come in all shapes and sizes. Canoes take a lot of gear and usually two or three people. Kayaks can be slimmed-down whitewater versions or beefy touring kayaks, sometimes called sea kayaks. There are different types of boats, such as rafts and dories used for running rivers. Whatever the case, you should be well familiarized with the skills needed for your particular boat.

The clothing you will need for canyon paddling is similar to the outdoor clothing described in Chapter 3. Consider wearing a special paddling jacket and pants that usually have tight cuffs that seal out water. This helps you stay warm and dry despite getting regularly sprayed by water. Another option is to wear a wet suit. Gloves protect your hands while paddling. Warm neoprene paddling or canyoneering shoes will keep your feet warm even when wet.

What type of paddling gear you need also depends on the type of boat you have. In general you will need paddles or oars specific to your boat. Always carry a spare in case you lose one to the water. Bailing gear, including a sponge, bailing cup, and hand pump, are essential.

A life vest is usually mandatory and can save your life. Use a Coast Guard–approved personal flotation device or life jacket. A Type 1, or off-shore, life vest is designed to turn most unconscious wearers face up. A Type 2, or near-shore, is less bulky and more comfortable but provides less flotation. In either case, make sure your vest fits well. Other types of personal flotation devices designed for paddling may be more comfortable but have less flotation.

A throw rope for rescue and repair kit with patch supplies are also essential. Remember, your patch kit and spare parts will be specific to your boat.

Table 4

ADDITIONAL EQUIPMENT FOR PADDLERS*

CLOTHING

Helmet

Hat that fits under helmet

Gloves, fingerless or full-fingered

Wet suit top and/or bottom

Paddle jacket and pants

Shoes, neoprene

PADDLE GEAR

Paddle(s), with spare

Paddle float (kayaks)

Personal Flotation Device (PFD)

Hand pump

Sponge and bailer cup

Throw rope

Spray skirt (for kayaks)

Bow and stern rope

Patch and repair kit (especially for rafts)

* This table lists items to take on a paddling trip in addition to those listed in Table 1 (pages 34–35), General Equipment for Canyoneering. Make sure all repair tools and parts are compatible with your brand of boat and accessories.

PACKING

With boat travel, you can usually take more gear than with backpacking or biking. Make sure everything is stowed in waterproof drybags. Taking a five-gallon bucket with a watertight lid is another option. Make sure you load your craft evenly from both side to side and bow to stern.

PLANNING A TRIP

When planning your paddle trip, use the route finding, weather reading, and other skills outlined in Part I of this book. Obstacles in water can be particularly treacherous, especially in rapids, swift water, and turbulent eddies; watch our for submerged rocks and logs. Make sure you have a good map and have scouted well. Check up on seasonal and yearly variations of water flow. Many rivers can only be run safely during certain times of year. Seasonal runoff can dramatically change stream levels and currents. Always start out on gentle water. Gradually work up to more advanced routes. Use a guide or partner who knows boating and the river.

There are basically two types of canyon paddling trips: downstream-only or up-and-back. Many river trips go downstream only, as the river is too swift to reasonably paddle up. A downstream float is much easier to paddle and takes less time. However, in most cases downstream trips require a car shuttle, one to drop you off at your put-in and another to pick you up downstream. Sometimes you may be able to leave a mountain bike at the takeout and bike back to a car you have left at the start of your trip, provided it isn't too far away.

An up-and-back is typically done in a canoe or kayak. It usually involves paddling upstream to a point, then turning around and floating back downstream. This saves a car shuttle and sometimes is the only way to see canyons that don't have vehicle access. Another benefit with an up-and-back trip is that you can turn around at any time in cases of poor weather, injury, or fatigued paddlers.

When planning time, keep in mind that it is much slower paddling upstream than downstream. Wide, smooth stretches will take much longer to get through than narrow, deep, swift sections. Portages around obstacles and breaks for side trips will add travel time as well. Likewise, a head wind can significantly slow the process. If you are drifting downstream, keep in mind that a car shuttle may take a half or full day at the beginning and end of a trip.

Also, make sure you are paddling in an area where boating is allowed, and that you have secured the proper permits. Many wilderness areas, national parks, and national forests have limitations as to where you can take watercraft. Access to water by vehicle for put-in and pull-out may be limited as well. Learn the rules before you go.

SAFETY

Always wear a life vest when in the water. Always wear a helmet and appropriate clothing—for example, a wet suit or paddling jacket in cold water, and gloves.

Scout rapids and watch for eddies and swift water. Swift water can be deep and narrow or wide and shallow.

If you fall into the water, float in the sitting position with your legs out in front of you. Keep your knees bent and use your hands and feet to fend off submerged boulders or logs. Try not to fight the current but rather swim to the bank that the river naturally flows to. Usually you can get to one bank or the other and find an eddy to slow down in.

LOW-IMPACT PADDLING

As with any type of canyoneering, follow the guidelines outlined in Chapter 1. Manage your waste correctly. Choose your campsites so that they minimize damage to the stream banks. Use care when entering and exiting your boat, and preserve the riparian environs. Secure your boat to the bank or pull it out of the water at the end of your trip.

Swiftwater roll, Idaho (photo © Erik Benner)

ON THE WATER

The water was cold and clear unlike the turbid chocolate waters of the undammed river from many years ago. Dawn peaked over the canyon cliffs and lit the sky in yellows and blues.

Our camp, 10 miles upstream from the Grand Canyon put-in, was on the river's edge, nestled amidst non-native tamarisk and blooming yucca. Our destination, Glen Canyon Dam, was upstream another 5 miles. From my tent I could see large dark holes in the rock cliff; later I figured out these were windows in the 5-mile-long tunnel that led trucks from the rim to the river during construction of the dam.

Paddling upstream was slow, but we made steady progress. Around the last bend we came to the huge monolith that plugs Glen Canyon, that backs up water and silt for miles in Lake Powell, that floods miles and miles and miles and miles of slot canyons. An unnatural, almost surreal, human-made eyesore. After two days of paddling in this wilderness, the concrete was anticlimactic; we had not reached a goal but a place to turn around.

Some thirty years ago Earth First! rallied here with a simulated crack in the middle of this dam. Now periodic flooding, letting down the water level behind the dam, and even partial removal of the dam are serious topics of discussion. Only months after our trip, there was to be a widely publicized flood of the Grand Canyon to revitalize the ecosystem that for so long missed the annual raging floods of nature.

For now I'm content in my little kayak, packed with provisions for another day or two. I am dwarfed by the dam, dwarfed by the vastness of river and rock, spinning around in an eddy and the current. I listen to the water roaring through the penstocks, stifling the canyon wren's song. Downstream the paddle will be much smoother, easier, shorter. In no time I'll be out of sight and earshot of the concrete cork. Back into the wilderness.

Infamous Glen Canyon Dam, viewed from kayak

APPENDIX 1

CANYONEERING CLASSIFICATIONS

GENERAL CANYONEERING LEVELS

Several canyoneering rating systems are being used today. Most of these are devised by guide companies and are not widely used. The following classification system is a general guide that is designed to give readers an idea of different levels of canyoneering.

CLASS 1: CANYON HIKING

Usually dry hiking on established trail

Routefinding straightforward using map and guidebook

Established campsites available

Hike to rim or backtracking straightforward

Evacuation of injured person straightforward

No major obstacles

CLASS 2: ADVANCED CANYON HIKING

Dry hiking with some river rock or uneven terrain

Streams easily crossed by ankle-deep wading or jumping

Straightforward routefinding with map and guidebook

Hike to rim or backtracking straightforward

Suitable campsite available

Evacuation of injured person straightforward

Obstacles such as boulders easily circumvented

CLASS 3: BASIC CANYONEERING

Prolonged water hiking and uneven terrain

Stream crossings utilize downed logs, rocks, or wading

Class 3 climbing such as scrambling over log jams may be necessary;
    may be able to circumvent

Navigation and routefinding skills necessary

Exits to rim sparse

Backtracking may be difficult

High and dry campsite more difficult to find

Risks of flash flood, cold injury, drowning, rockfall more prominent

Evacuation of injured person difficult

CLASS 4: ADVANCED CANYONEERING

Difficult hiking, multiple stream crossings, prolonged water walking

Tight narrows may require taking off pack

Plunge pools and wall-to-wall water may require swimming or deep wading

Class 4 climbing with hand lines necessary in some places

Circumventing log jams or chockstones may not be possible; may require bouldering skills

Navigation and routefinding skills necessary

Exit to rim difficult to find

Backtracking difficult

High and dry campsite difficult to find

Risks of injury and environmental hazards more significant

Evacuation of injured person difficult

CLASS 5: TECHNICAL CANYONEERING

Difficult and slow hiking, multiple stream crossings, prolonged water walking

Tight narrows may require taking off pack

Plunge pools and wall-to-wall water may require swimming across

Class 5 climbing with hand lines and bouldering skills necessary

Rappels necessary for dry pouroffs or waterfalls; anchors such as trees available

Navigation and routefinding very difficult and time-consuming

Exit to rim more difficult to find

Backtracking very difficult

High and dry campsite rare

Combining of skills such as rappelling into a pool or in a waterfall needed frequently

Risks of injury and environmental hazards significant; some risk of death

Evacuation of injured person extremely difficult without outside help

CLASS 6: ADVANCED TECHNICAL CANYONEERING

Extremely difficult, slow hiking and water walking

Tight narrows may require taking off pack

Long swims and difficult, technical stream crossings with hand lines

Hand lines necessary for down climbing

Rappels frequent; anchors not readily available

Class 5.7 and above rock climbing

Navigation and routefinding very difficult and time-consuming

Secure, high and dry campsite may not be available

Risks of injury and environmental hazards constant, and risk of death
significant

Combining of skills, such as rappelling into a pool or in a waterfall, frequently necessary

Exit to rim impossible

Backtracking impossible without fixed lines

Evacuation of injured person nearly impossible without help

AMERICAN CANYONEERING ASSOCIATION RATING SYSTEM

The American Canyoneering Association recently developed a rating system for canyoneering. It is similar to that used in Europe by the European Canyoneering Commission. The system uses four sets of variables, either numbers, letters, or both. Ratings refer to canyons in their usual conditions during canyoneering season for that particular area. Seasonal and yearly changes are not accounted for.

ROPE WORK (FIRST VARIABLE, ONE DIGIT)

1 No rope required for normal descent, although a rope should always be carried for emergencies.

2 Only single-pitch rappels required for normal descent.

3 Complex rope work and/or multi-pitch rappels required.

WATER (SECOND VARIABLE, TWO DIGITS)

Volume/Current

1 No water (or very, very little water).

2 Water with no current or light current.

3 Water with strong current.

Swimming

A No swimming required.

B Short and/or infrequent swimming required.

C Long and/or frequent swimming required.

TIME (THIRD VARIABLE, ONE OR TWO DIGITS)

Average time (based on climbing grades)

1 1 to 3 hours.

2 3 to 4 hours.

3 3 to 6 hours.

4 Full day.

5   1 to 2 days.

6   Multiple days in the canyon.

EXITS (CLASS 4 OR EASIER)

R   Some sections with limited exits.

X   Long sections with no possible exits.

CLIMBING (FOURTH VARIABLE, OPTIONAL)

Classes of terrain based on the system developed by the Sierra Club; technical climbing (Class 5) is defined more precisely using the Yosemite Decimal System.

1   Walking on level ground.

2   Hiking on or off trail with some elevation change.

3   Scrambling; may require hands for balance and support.

4   Easy vertical or near vertical climbing

5   Technical climbing; further broken down by Yosemite Decimal System from 5.0 to 5.14

6   Aid climbing.

Note: It can be assumed that all canyons will involve some terrain up to and including Class 3. If terrain is encountered that is rated Class 4 or higher, the rating should be appended with a V to indicate the difficulty of the hardest vertical climbing (or down climbing) moves (i.e. V4 for 4th class, V5.4 for 5.4 climbing moves, V6 for aid climbing).

EXAMPLES OF RATINGS

2-2A-4

2   Descent will require single-pitch rappel(s).

2A   Canyon normally has water with no current or light current, but does not require swimming. Water can be avoided or only requires wading.

4   An average group will take a full day to descend. Because the time value (4) is not followed by an R or X, it can be assumed that there will be numerous exit opportunities. There is no V appended to the end of the rating, so it can be assumed that the terrain encountered will be Class 3 or easier.

3-2C-3X-V5.6

3   Descent will require complex rope work and/or multi-pitch rappel(s).

2C   Canyon normally has water with no current or light current and will require long and/or frequent swims.

3X   An average group will take 4 to 6 hours to descend. The X indicates that there are long sections where exit is not possible.

V5.6   Climbing will be encountered with moves rated up to 5.6 difficulty.

This rating system was developed by Rich Carlson and adopted by the American Canyoneering Association. Used with permission.

APPENDIX 2

ADDITIONAL RESOURCES

There are resources available to help you as you begin your canyoneering adventures. Because canyoneering is a relatively new sport, more and more resources will become available. This is a basic list to get you started.

BOOKS

Numerous books provide in-depth discussions of outdoor equipment, backpacking, caving, mountaineering, rappelling, rock climbing, weather reading, and navigation. The following is a basic list of titles that may be useful.

Burns, B. and M. Burns. *Wilderness Navigation: Finding Your Way Using Map, Compass, Altimeter, & GPS.* Seattle: The Mountaineers, 1999.

Carline, J. D., M. J. Lentz and S. C. MacDonald. *Mountaineering First Aid,* 4th ed. Seattle: The Mountaineers, 1996.

Craighead, F. C. et al. *How to Survive on Land and Sea,* 4th ed. Annapolis, Md: Naval Institute Press, 1984.

Darvill, F.T. *Mountaineering Medicine: A Wilderness Medical Guide,* 14th ed. Berkeley, Calif.: Wilderness Press, 1998.

Davenport, G. J. *Wilderness Survival.* Mechanicsburgh, Penn.: Stackpole Books, 1998.

Fasulo, D. J. *Self-Rescue.* Evergreen, Colo.: Chockstone Press, 1996.

Fleming, J. *Staying Found: Complete Map and Compass Handbook,* 2nd ed. Seattle: The Mountaineers, 1994.

Fletcher, C. *The Complete Walker III.* New York: Knopf, 1984.

Frank, J. and D. Patterson. *CMC Rappel Manual,* 2nd ed. Santa Barbara, Calif.: CMC Rescue, 1997.

Ganci, D. *Desert Hiking*. Berkeley, Calif.: Wilderness Press, 1993.

Graydon D., editor. *Mountaineering: The Freedom of the Hills,* 6th ed. Seattle: The Mountaineers, 1998.

McGivney, A. *Leave No Trace: A Guide to the New Wilderness Etiquette.* Seattle: The Mountaineers, 1998.

Hodgson, M. *The Compass and Map Navigator: A Complete Guide to Staying Found.* Merriville, Ind.: ICS Books, 1997.

Martin, T. *Rappelling,* 2nd ed. Mt. Sterling, Ky.: Search, 1995.

Powers, P. *NOLS Wilderness Mountaineering*. Mechanicsburg, Penn.: Stackpole Books, 1993.

Rea, G. T. *Caving Basics: A Comprehensive Guide for Beginning Cavers,* 3rd ed. Huntsville, Ala.: National Speleological Society, 1992.

Schimelpfenig, T. and L. Lindsey. *NOLS Wilderness First Aid.* Mechanicsburg, Penn.: Stackpole Books, 1992.

Townsend, C. *The Backpacker's Handbook,* 2nd ed. Camden, Me.: Ragged Mountain Press, 1997.

Wilkerson, J. A., editor. *Medicine for Mountaineering and Other Wilderness Activities,* 4th ed. Seattle: The Mountaineers, 1992.

Woodmencey, J. Reading. *Weather: Where Will You Be When The Storm Hits?* Helena, Mont.: Falcon, 1998.

INTERNET SITES

The Internet is an immense source for canyoneering information. You can find useful information if you know how to surf the net with efficiency. Search engines are a great starting point. Outfitters and guide services, including those overseas, often have websites. Mail-order companies have gear available online as well.

The few websites worth a regular visit are weather sites. Agencies such as the US Forest Service or National Park Service can also be useful. Below are the home pages for a few selected sites.

CANYONEERING

American Canyoneering Association: *www.canyoneering.net*

Commission Européen de Canyon: *www.cec.canyoning.org*

WEATHER

National Weather Service (United States): *www.nws.noaa.gov*

Weather Net (North America): *cirrus.sprl.umich.edu/wxnet*

Canada Meteorological Center: *www.cmc.doe.ca*

National Weather Service (United Kingdom): *www.meto.govt.uk*
Weather Channel (United States and World): *www.weather.com*
World Meteorological Association: *www.wmo.ch*
UNITED STATES GOVERNMENT AGENCIES
Bureau of Land Management: *www.blm.gov*
United States Forest Service: *www.fs.fed.us*
National Park Service: *www.nps.gov*

OTHER RESOURCES

Check your local telephone book or a climbing, mountaineering, or camping store for additional up-to-date information on weather, road conditions, guided trips, or courses covering backcountry first aid, CPR, mountaineering, or climbing.

APPENDIX 3

GLOSSARY

abyss a dark deep hole; a chasm

aiders loops of webbing used with ascenders to climb a rope

altimeter a barometer that calculates elevation

arroyo a mostly dry canyon that is cut by occasional water flow

ascender a mechanical device used to climb a rope

ascending the act of climbing up a rope

aspect the direction in which a slope faces

bench (also **ledge**) a flat part of a canyon wall or rim

bisection technique that identifies a position on a map by using one field bearing in the canyon to locate position

bivouac (also **bivy**) a small temporary camp or unscheduled overnight

bivouac sack a one-person tube shelter

bollard a stack of rocks that serves as a rappel anchor

bouldering rock climbing low to the ground without the protection of a rope

bowline a knot used to tie a loop at one end of a rope

box canyon a canyon that ends with a cliff or falls with no easy escape route

brook (also **creek**) a small shallow tributary

cairn a short stack of rocks used to mark a trail

canyoneering (also **canyoning**) traveling through canyons

carabiner a large snap link

chasm (also **abyss**) a dark deep hole

chock a climbing protection tool that one wedges into a crack

chockstone a rock or boulder lodged in a canyon or crack

chimney (also **counterforce** or **stem**) a counterforce climbing maneuver using legs and arms in opposition

citronella a natural, citrus-based insect repellent

Colorado tick fever an illness caused by a virus transmitted by the wood tick

cordlette (also **reapschnur** or **retrieval cord**) a retrieval cord used for an advanced rappel technique

couloir a narrow, steep chute

coccidiomycosis a microorganism found in dust

counterforce (also **stem** or **chimney**) a climbing technique using pressure in opposite directions that can be used when no obvious holds are available

crack a narrow opening in a canyon wall

creek (also **brook**) a small, shallow tributary

declination the local variance between magnetic north and true north measured east or west in degrees

DEET the chemical N,N-diethylmetatoluamide used in some insect repellents

dew point the temperature at which air is at 100 percent relative humidity

double fisherman's (also **grapevine**) a knot used to tie a cord into a sling or for tying two ropes together

dryfalls waterfalls that have no running water part of the year

dry suit a suit made from rubberized nylon with gaskets at the neck and cuffs to seal out water

dynamic rope an elastic rope designed for lead climbing

edging a climbing technique that involves stepping on a ledge using the edge of one's shoe

8 (also **figure 8**) a rappel/belay device

expansion bolt a bolt drilled and hammered into the rock to provide an anchor for climbing or rappelling

Farmer John a long-legged, sleeveless wet suit

field bearing a compass reading taken from landmarks in the field based on magnetic north

figure 8 (also **8**) 1) a knot used to tie a rope into a harness, to tie a loop in the end of a rope, or to tie two ropes together; 2) a metal rappel/belay device

flash flood a sudden flood from rain

freestanding tent a tent that does not need guy lines or stakes to be set up

freezing level the elevation at which the air is 32°F (0°C)

front the meeting point of one pressure system or air mass with another

frostbite a skin injury caused by freezing

frostnip a minor skin irritation from the cold

giardia a microscopic parasite often found in mountain streams

global positioning system (GPS) An electronic device that determines position in longitude and latitude by using navigational satellites

gorge a large canyon

grapevine (also **double fisherman's**) a knot used to tie a cord into a sling or for tying two ropes together

gully a small canyon or channel cut by running water that is usually smaller than a gulch or ravine

gulch a small canyon or channel cut by running water; similar to a ravine

guy line a cord to tie a tent to the ground using a stake at one end

hand line a piece of rope or webbing used for extra support when climbing or crossing a stream

hanta virus a virus transmitted by the deer mouse and other small mammal

heat exhaustion an illness caused by rising body temperature and dehydration

heat stroke shock caused from rising body temperature and dehydration

histoplasmosis a microorganism found in caves contaminated with bat guano and bird droppings

humidity the amount of water vapor in the air

hypothermia an illness caused by lowering of the body core temperature

isobars lines that connect similar pressures on a weather map

jamming a climbing technique that involves wedging a foot or hand into a crack

jumaring an ascending technique

kernmantle modern climbing rope construction using a core and outer sheath

Klemheist a knot tied in webbing to provide friction used for ascending rope

latitude (also **meridian**) the north-south lines on a map used with longitude to identify one's position on earth

ledge (also **bench**) a flat part of a canyon wall or rim

lock link a steel screw chain link

longitude the east-west lines on a map used with latitude to identify one's position on earth

lyme disease an illness caused by a bacteria transmitted by the deer tick, the Western black-legged tick, and others

macrobiotic soil the collection of algae, lichen, and moss covering the desert sand that provides a seedbed for plants, holds moisture, and fixes nitrogen

magnetic north the bearing to which all compasses point, which is different from true north

mantel a climbing maneuver preformed by applying downward pressure on a ledge to lift oneself upon it

map bearing a compass reading taken from a map based on true north

meteorology the study of weather

meridian the north-south lines used with longitude to identify one's position on earth

Münter hitch a knot tied on a carabiner used to rappel down a rope

narrows the sections of canyon with walls close together

National Climbing Classification System a system for grading difficulty of technical alpine climbs in the United States

objective risks risks that occur from being in the wrong place at the wrong time

oral rehydration solution electrolyte and water mixture that provides hydration and replacement of salts

orienteering map and compass navigation

panniers bike packs designed to fit over a front or rear rack

permethrin a chemical in some tick repellents

petroglyph a carved or pounded image on rock

pictograph a drawing or painting on rock

piton a metal spike hammered into rock for protection or to provide an anchor for climbing or rappelling

plunge pool a collection of water at the bottom of a waterfall

pressure the amount of force a mass of air exerts on the surrounding atmosphere

primary survey evaluation of an injured person by checking the airway, breathing, and circulation, in that order

prusik a knot in cord that provides friction to ascend a rope

punchbowl a large plunge pool

rabies an illness caused by a virus; usually contracted when a human is bitten by an infected animal

rappel ring a metal ring used in a rappel station to allow the rope to easily slide though webbing

rappelling sliding down a rope

ravine a small canyon or channel cut by running water; similar to a gulch

reapschnur (also **cordlette** or **retrieval cord**) a retrieval cord used for advanced rappel techniques

relative humidity the percentage of water vapor in the air compared to the point of complete saturation

retrieval cord (also **cordlette** or **reapschnur**) a cord used to remove webbing or a rappel rope

ridge high-pressure isobars on a weather map

rim rocked after climbing up on a boulder or ledge, being unable to execute the more difficult climb down

rock climbing ascending or descending sections of rock too steep to simply hike

Rocky Mountain spotted fever an illness caused by bacteria, often transmitted by the wood tick and the American dog tick

runner webbing or cord tied in a loop

sag wagon a support vehicle for bikers that carries equipment

scree small, loose stones on a hill side

secondary survey a more detailed head-to-toe assessment of an injured person

shakedown a practice session

shorty a short-legged, short-sleeved wet suit

sling webbing tied in a loop

slot canyon a narrow canyon, sometimes just wide enough to squeeze through

smear a climbing maneuver that relies on the friction of the shoe sole to keep one on the rock

spotter a person who can slow or prevent a fall for one who is bouldering

sprain an injury to joint ligaments and tendons

spring-loaded camming device a climbing protection tool that one retracts, places in a crack, then expands

static rope low-stretch rope designed for rappels

stem using feet or hands in opposition against two walls for climbing

strain a stretch or minor tear to a muscle

subjective risks risks one minimizes with safety techniques and equipment

sun protection factor (SPF) the rating of a sunscreen with respect to the amount of protection it gives you from the sun's rays

tableau a collection of artifacts such as arrowheads and pottery shards

talus loose rock on a hillside

topographic map a map with lines that connect points of the same elevation

torrent a swift, turbulent flow of water

trench foot damage to feet from prolonged exposure to cold water

triangulation identifying your position on a map by using the intersection of two field bearings

trough low-pressure isobars on a weather map

true north the direction, pointing to the north pole, on which all maps are based

tube a rappel/belay device

Union Internationale des Associations d'Alpinisme (UIAA) an organization that tests, rates, and certifies climbing gear

water intoxication a type of heat illness marked by lack of salt intake and profuse sweating

water knot a knot used to tie a piece of webbing into a sling or to tie two pieces of webbing together

wind chill the ambient temperature adjusted for the cooling effects of wind on bare skin

Yosemite Decimal System a system used in the United States to rate difficulty of climbs, class 5.1–5.14

Index

(table, photo, or fig. in parentheses)

ABOUT THE AUTHOR

Christopher Van Tilburg is an emergency and wilderness medicine physician who frequently writes and lectures on backcountry safety. He is the adventure sports editor for *Wilderness Medicine Letter*, a former medical columnist for *Wind Tracks* magazine, and a member of Portland Mountain Rescue, the Wilderness Medical Society, and numerous conservation organizations. His first book, *Backcountry Snowboarding*, was published by The Mountaineers Books in 1998. Chris has hiked, biked, climbed, paddled, driven,  and rappelled through Pacific Northwest lava canyons, Colorado Plateau red rock slots, Mexican arroyos, and limestone narrows in the Alps. He lives in White Salmon, Washington, in the Columbia River Gorge.

THE MOUNTAINEERS, founded in 1906, is a nonprofit outdoor activity and conservation club, whose mission is "to explore, study, preserve, and enjoy the natural beauty of the outdoors. . . . " Based in Seattle, Washington, the club is now the third-largest such organization in the United States, with 15,000 members and five branches throughout Washington State.

The Mountaineers sponsors both classes and year-round outdoor activities in the Pacific Northwest, which include hiking, mountain climbing, ski-touring, snowshoeing, bicycling, camping, kayaking and canoeing, nature study, sailing, and adventure travel. The club's conservation division supports environmental causes through educational activities, sponsoring legislation, and presenting informational programs. All club activities are led by skilled, experienced volunteers, who are dedicated to promoting safe and responsible enjoyment and preservation of the outdoors.

If you would like to participate in these organized outdoor activities or the club's programs, consider a membership in The Mountaineers. For information and an application, write or call The Mountaineers, Club Headquarters, 300 Third Avenue West, Seattle, Washington 98119; (206) 284-6310.

The Mountaineers Books, an active, nonprofit publishing program of the club, produces guidebooks, instructional texts, historical works, natural history guides, and works on environmental conservation. All books produced by The Mountaineers are aimed at fulfilling the club's mission.

Send or call for our catalog of more than 300 outdoor titles

The Mountaineers Books
1001 SW Klickitat Way, Suite 201
Seattle, WA 98134
800-553-4453
mbooks@mountaineers.org
www.mountaineersbooks.org

Other titles you may enjoy from The Mountaineers:

MOUNTAINEERING: The Freedom of the Hills, Sixth Edition, *The Mountaineers*
The completely revised and expanded edition of the best-selling mountaineering "how-to" book of all time—required reading for all climbers.

CONDITIONING FOR OUTDOOR FITNESS: A Comprehensive Training Guide, *David Musnick, M.D., and Mark Pierce, A.T.C.*
The most comprehensive guide to conditioning, fitness, and training for all outdoor activities. Chapters on specific sports including hiking, climbing, biking, paddling, and skiing, with information on cross-training.

BACKCOUNTRY MEDICAL GUIDE, Second Edition, *Peter Steele*
A pocket-sized, comprehensive medical guide for backcountry emergencies, this book emphasizes techniques for situations when the victim is far from help. Includes a detailed list of medications and what to include in your first-aid kit.

WILDERNESS NAVIGATION: Finding Your Way Using Map, Compass, Altimeter, and GPS, *Bob Burns and Mike Burns*
A guide to navigating both on the trail and off in the backcountry. Includes the most reliable and easy-to-learn methods of navigation yet devised.

EVERYDAY WISDOM: 1001 Expert Tips for Hikers, *Karen Berger*
Expert tips and tricks for hikers and backpackers selected from one of the most popular *BACKPACKER* Magazine columns. A one-stop, easy-to-use collection of practical advice, time-saving tips, problem-solving techniques and brilliant improvisations to show hikers how to make their way, and make do, in the backcountry.

MOUNTAIN BIKE ADVENTURES IN™ Series:
Complete guides to off-road cycling.
MOUNTAIN BIKE ADVENTURES IN™ SOUTHWEST BRITISH COLUMBIA: 50 Rides
MOUNTAIN BIKE ADVENTURES IN™ THE FOUR CORNERS REGION
MOUNTAIN BIKE ADVENTURES IN™ THE NORTHERN ROCKIES
MOUNTAIN BIKE ADVENTURES IN™ WASHINGTON'S NORTH CASCADES & OLYMPICS, Second Edition
MOUNTAIN BIKE ADVENTURES IN™ WASHINGTON'S SOUTH CASCADES & PUGET SOUND, Second Edition

BACKCOUNTRY SNOWBOARDING, *Christopher Van Tilburg*
This is the first book to introduce snowboarders to the techniques and concerns of backcountry snowboarding. Fully illustrated with color photos throughout, this book covers everything the advanced freerider needs for safe ascent and descent.